Crop Nutrient Uptake and Efficiency

Crop Nutrient Uptake and Efficiency

Edited by:

Kiran

Delve Publishing

www.delvepublishing.com

Crop Nutrient Uptake and Efficiency

Kiran

Delve Publishing
224 Shoreacres Road
Burlington, ON L7L 2H2
Canada
www.delvepublishing.com
Email: orders@arclereducation.com

ISBN: 978-1-77469-033-8 (Hardcover)

ABOUT THE EDITOR

Dr. Kiran received her undergraduate education from Bangalore University, B.Sc. in Biotechnology. Then she did postgraduation in M.Sc Ag. from Allahabad University, where she got a gold medal. Then she did her Ph.D (Ag.) in Plant Breeding and Genetics from CSKHPKV, Palampur. She has a brilliant academic record throughout. She worked at G.B. Pant University of Agriculture and Technology, Pantnagar and GKU,Bathinda .She has a number of books ,research papers to her credit. Her research interests range from Plant Breeding & Genetics, Seed Science & Technology, Biotechnology, Bioinformatics and Plant Tissue Culture. She has voluntarily worked as a coordinator, society for social development, as a community development trainer, Raipur and as a public support executive, Navjeeven foundation, Kangra HP, India. Presently Dr Kiran is working as Assistant Scientist (Genetics and Plant Breeding) in Chaudhary Sarwan Kumar Himachal Pradesh Krishi Vishvavidyalaya, Palampur, India.

TABLE OF CONTENTS

List of Figures .... *xi*
Preface .... *xv*

Introduction 1

Chapter 1 Introduction to Crop Nutrients 3
1.1 Introduction 4
1.2 Nutrient Uptake 7
1.3 Nutrient Use Efficiency 8
1.4 Nutrient Dynamics 11
1.5 Environmental Issues 20

Chapter 2 Essential Plant Nutrients 25
2.1 Introduction 26
2.2 Nitrogen 30
2.3 Phosphorus 33
2.4 Potassium 35
2.5 Calcium 36
2.6 Magnesium 37
2.7 Sulfur 38
2.8 Zinc 40
2.9 Copper 42
2.10 Iron 43
2.11 Manganese 44
2.12 Nickel 46
2.13 Chlorine 46
2.14 Molybdenum 47
2.15 Boron 48

Chapter 3 Nutrient Efficiency and Management 51
3.1 Introduction 52
3.2 Physiology 54
3.3 Nutrient Use Efficiency 57
3.4 Nue For Macronutrients 63

Chapter 4 Bacteria in Plant Nutrient Management 71
4.1 Introduction 72
4.2 Mechanism 74
4.3 Zinc Absorption 82
4.4 Uptake of Phosphate 85
4.5 Sulfur Oxidizing Bacteria 86
4.6 Siderophores 89
4.7 Phytohormones 92

Chapter 5 Molecular Basis of Nutrient Efficiency 97
5.1 Introduction 98
5.2 Root Systems 98
5.3 Rhizosphere 103
5.4 Canopy Physiology 108
5.5 Remobilization of the Nutrients 109
5.6 Grain Composition 111
5.7 Genetic Improvement 112
5.8 Molecular Genetics 113

Chapter 6 Metabolic Networks 117
6.1 Introduction 118
6.2 Photorespiration 119
6.3 Measurement of Metabolites 122
6.4 Enzymes 127
6.5 Isotopes 128
6.6 Topology 128
6.7 Dynamic Flux Analysis 130
6.8 Primary Metabolites 130

Chapter 7 Nutrient Deficiencies **137**
7.1 Introduction 138
7.2 Maize 139
7.3 Rice 143
7.4 Sorghum 146
7.5 Genome Sequencing 151
7.6 Phosphorus Uptake 153
7.7 Improved Crop Quality 154

Chapter 8 Nanotechnology in Crop Nutrients Studies **157**
8.1 Introduction 158
8.2 Biomass and Yield 169
8.3 Nano Fertilizers 173
8.4 Nanopesticides 178
8.5 Applications of Nanotechnology 184
8.6 Management of Pests 186
8.7 Environmental Problems and Challenges 191
8.8 Indirect Measurement Techniques 194
8.9 Conclusion 196

References **197**

Index **223**

LIST OF FIGURES

Figure 1.1 Crop nutrients
Figure 1.2 Crop nutrient uptake
Figure 1.3 Nutrient use efficiency
Figure 1.4 Plant nutrient dynamics
Figure 1.5 Plant biotic stresses
Figure 1.6 Plant abiotic stresses
Figure 1.7 Integrated nutrient management
Figure 1.8 Green manure
Figure 1.9 Farmyard manure
Figure 1.10 Biofertilizers
Figure 1.11 Groundwater contamination
Figure 2.1 Essential plant nutrients
Figure 2.2 Transporter proteins
Figure 2.3 Nitrogen cycle
Figure 2.4 Phosphorous deficiency
Figure 2.5 Potassium deficiencyon leaves
Figure 2.6 Calcium deficiency
Figure 2.7 Magnesium deficiency
Figure 2.8 Sulfur deficiency
Figure 2.9 Zinc deficiency
Figure 2.10 Copper deficiency
Figure 2.11 Iron deficiency
Figure 2.12 Manganese deficiency
Figure 2.13 Molybdenum deficiency
Figure 2.14 Boron deficiency
Figure 3.1 Fertigation
Figure 3.2 Benefits of VAM fungi
Figure 3.3 QTL

Figure 3.4 Recombinant inbred lines
Figure 3.5 Linkage disequilibrium
Figure 3.6 Sulfate transport proteins
Figure 3.7 Plant hormones
Figure 4.1 Plant growth-promoting rhizobacteria
Figure 4.2 Nitrogenase mechanism
Figure 4.3 Zinc absorption
Figure 4.4 Siderophores
Figure 4.5 Biosynthesis of IAA
Figure 5.1 Root system
Figure 5.2 Lateral root meristem
Figure 5.3 P uptake and roots
Figure 5.4 Rhizosphere zone activity
Figure 6.1 Plant metabolic networks
Figure 6.2 Photorespiration
Figure 6.3 NMR technique
Figure 7.1 Plant nutrient deficiencies
Figure 7.2 Genomic DNA
Figure 7.3 NGS technologies
Figure 8.1 Nanotechnology in Agriculture
Figure 8.2 Nanofertilizers
Figure 8.3 Nanocapsules
Figure 8.4 Nanoclays
Figure 8.5 Nanoparticles
Figure 8.6 Nanocontainers
Figure 8.7 Gold nanoparticles
Figure 8.8 Penetration of nanoparticles
Figure 8.9 Secondary metabolites
Figure 8.10 Nanoporous zeolites
Figure 8.11 Phosphorus nanofertilizers
Figure 8.12 Nanocomposites
Figure 8.13 Nanopesticides
Figure 8.14 Nanosuspensions
Figure 8.15 Nanofarming

Figure 8.16 Hydroxyfullerenes
Figure 8.17 Nanosensors
Figure 8.18 Nanodisks

PREFACE

Crops were considered vital source of human nutrition for ages. It is obvious that in the coming years, there will be an increased demand for food products and allied materials. Humans are being pushed to adopt sustainable methods to reach the expectations of the population. It is sensed that there is a need for combined attempts of scientists, plant breeders, farmers, agronomists, and nutritionists to achieve the expected goal of sufficient food for all. Food security has been a burning problem for decades. It is always important to practice effective nutrient management strategies to combat the present situation of dwindling crop yields. It is important to know about the nutrients that support plant growth and development. Agroecosystems are tagged as dynamic entities. This dynamic property can be attributed to the productivity, turnover of the fertilizers and nutrients, genetic stocks, and the crop types being used. These entities supply food for animals and humans and as the demand gets enhanced, there is a need for crop intensification strategies. Due to this intensification process, many agroecosystems were formed.

The plant kingdom can be considered as a primary source of both life and adhered economic development. Many fields like feedstocks, fuels, pharmaceuticals, and foods are intertwined with plants. It is important to understand the metabolic networks that can hold the integrated reactions occurring in the plant systems. The study of plant metabolism can be a challenging task and it needs to be supported by quantitative data analysis. We may need computational methods to arrive at a conclusion as far as the plant metabolic networks are concerned. The signal transduction process can be important in understanding the overall cellar activities that occur in the plant systems. Signal transduction can be important in plant physiology and development. We can understand the reactions of the plant towards environmental conditions. Genomic sequencing is one of several important methods that can help us understand the role of nutrients in crop development.

The plant nutrients can be considered vital in several aspects of a plant's life. Major nutrients will be absorbed from the atmosphere and some of the nutrients will be absorbed from the soils. The nutrients that are limited in the soils can be supplied in the form of fertilizers and manures. Every nutrient has a vital role in plant physiology. Abiotic and biotic stresses can be important as they bring considerable changes in plant physiology and development. This book deals with nutrients, metabolic networks and several other aspects of crop nutrient physiology and nutrient uptake. Much effort was made to make the entire book as simple as possible so that any reader who has the zeal in knowing the crop nutrient physiology can understand.

Introduction

Crop nutrients plays an important role in plant physiology and the macro and micronutrients can exhibit their presence whilst the growth and development of plants. The availability of cultivars and water can define the plant response to the soil conditions. It is important to determine the nutrient concentrations of the soil solution. The utilization of the nutrients, translocation, absorption, and availability of the nutrients can be considered important in crop physiology studies. The macronutrients can be classified into structural nutrients, primary nutrients, and secondary nutrients. The micronutrients are required by the plants in small proportions. The plant nutrients can be present in the relatively thin soil layer. Most of the plant nutrients are important to produce foods. The crop nutrients that are present in the soils may be due to the transformation of the rocks into minute soil materials. Some of the factors that define the action of the crop nutrients are parent material, topography, vegetation, climate, and gravity. Sustainable crop production is important so that the plant and microbial communities can be balanced. The microbes helps in improving the status of the soils so that fertility can get sustained.

In the primary chapters in this book, the concept of crop nutrients was discussed with special emphasis on classification and water availability in the soil solution. In the second chapter, the role of essential plant nutrients was discussed in detail. The third chapter deals with nutrient efficiency and other associated concepts that illustrate the actual role of the essential nutrients in plant growth and development. In the fourth chapter, the role of microbes like bacteria is explained along with important concepts like biological nitrogen fixation. In the fifth chapter, the molecular biology of integrated nutrient management and their role in defining plant physiology is explained in detail with relevant examples. In the sixth chapter, the metabolic networks that are present in the plant systems re explained with a special focus on nutrients absorbed by the plants. In the seventh chapter, the nutrient deficiencies and related plant pathology is explained.

This book can be used as an introductory course for both undergraduates and postgraduate students who would like to pursue a course in life and agricultural sciences. Those who are interested in the science of crops can use this as a handbook.

Chapter 1

Introduction to Crop Nutrients

CONTENTS

1.1 Introduction.. 4
1.2 Nutrient Uptake.. 7
1.3 Nutrient Use Efficiency 8
1.4 Nutrient Dynamics 11
1.5 Environmental Issues 20

1.1 INTRODUCTION

The role of minerals in the growth of the crop species is important. Mineral nutrition, crop cultivar type and water availability are important factor in agriculture. There are other factors like socio-economic conditions, weeds, insects, and diseases that are to be studied and observed by the farmers. Due to this knowledge gained, the farmers will be able to enhance crop productivity. The concentration of the nutrients can indicate the fertility of the soils. The green revolution has improved several aspects of agriculture in Asia and portions of Latin America. The crop yields were linked with the fertilizer application and it was estimated that about 60% of the yield was dependent on fertilizers especially in England and the United States (Stewart et al., 2005). Most of the crop yields in the 20th century were associated with the application of fertilizers (Baligar et al., 2001). In China, there was a 60% increase in the crop yields and it was attributed to the fertilizers (Baligar et al., 2001). In Brazil, there was a significant increase in the grain size of the rice on the application of N and P fertilizers. Several improvements were noticed in the oxisols and inceptisols in Brazil. The recovery of nitrogen in the cereals was decreased considerably. Due to reduced recovery efficiencies, the grain production of wheat, rice and other crops was reduced. Most of the agricultural soils are deficit in both micronutrients and macronutrients. It was stated by many researchers that the acidity can lower crop productivity.

The micronutrient deficiency can be due to intensive cropping practices, increased production demands that force to use marginal soils, enhanced fertilizer usage, reduced use of natural manures and other imbalances. The chemical fertilizers have affected most of the crop yields. According to the Food and Agricultural Organization, we may need 200 million tons of fertilizers per year and this might increase after the year 2030 (FAO, 2000). Euphrates and Tigris valleys were enriched with alluvial soils that supported many crops in 2500 BC (Russell., 1985). Until 200 BC, Romans and Greeks contributed several concepts of plant nutrition and soil fertility. Carl Sprengel proposed that the plants need elements for their development. The atmospheric nitrogen fixation by the legumes was considered as an important observation in plant science. Plants can absorb certain nutrients against energy gradients and this was named metabolic absorption or in some cases, we can call it as active absorption. The fertilizers must be added in limited concentrations and after a certain point, there will be an excess accumulation of the fertilizers that can bring down the crop yields. Von Liebig had proposed that the plant's nutritional status could serve as an indicator of the plant's resistance against diseases.

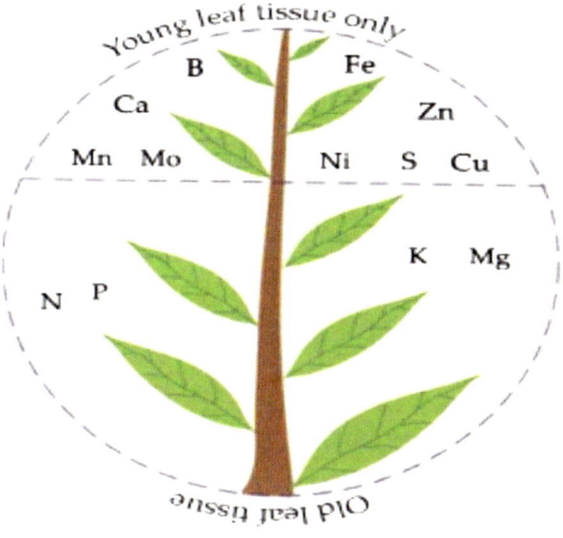

Figure 1.1: Crop nutrients.

Source: https://www.no-tillfarmer.com/articles/5829-identifying-nutrient-deficiency-symptoms-in-field-crops

The plants need a minimum of 17 elements and this number varies with species and these nutrients can be called essential nutrients. These nutrients (Figure 1.1) are important for the optimal growth of the plants. They are carbon-C, nickel-Ni, chlorine-Cl, molybdenum-Mo, boron-B, manganese-Mn, iron-Fe, copper-Cu, zinc-Zn, sulfur-S, magnesium-Mg, calcium-Ca, potassium-K, phosphorus-P, nitrogen-N, oxygen-O and hydrogen-H. The essential nutrients are those that are needed for plants to metabolize. If any nutrient is found as an intrinsic component in the plant systems and in its absence, a plant presents anomalies, then we can call it an essential nutrient. The plants can absorb carbon, hydrogen and oxygen from the water and air. For some of the nutrients other than C, H and O, most of the nutrients are absorbed from the soils. Each nutrient helps in the functioning of the plant cells and the unavailability of one nutrient can cause abnormal and impaired metabolism. The nutrients needed by the plants are classified into macronutrients and micronutrients. The macronutrients must be supplied to plants in large amounts. The micronutrients are needed in minor quantities. The need for nutrients in the crop growth can be based on soil biology, climatic conditions, type of soils, genotypes, crop species and level of yield.

The important factors that determine the nutrient requirements are climatic factors and soil. The nutrient deficiencies in the crop plants can occur due to soil erosion. Some of the nutrients are applied at low rates, water deficiencies, weeds, insects, immobilization, and other problems can cause nutrient deficiencies in the crop plants. There will be unfavorable conditions, imbalances, and elemental toxicity. The nutritional disorders are common in most crop plants all over the world. Some of the plant cultivars can present nutritional deficiencies compared to the wild varieties. The nutrient disorders can be identified by plant tissue tests, soil tests and visual symptoms. The soil test is considered an effective method in detecting nutritional disorders. It is important to follow sustainable agriculture where the inputs and outputs of the nutrients will be balanced. It is important to restore the nutrients that were lost in the crop harvest and crop rotation must be ensured. If the nutrients are accumulated in excess in the soils, the leaching process due to run-off can produce serious problems like eutrophication. The productivity in agriculture can be linked with the uptake and availability of nutrients. The availability of the nutrients can be associated with many processes in the soil-water-plant continuum. Some of the processes that are to be considered are the utilization of nutrients by plants, nutrient transport, and absorption of the nutrients. Most of these processes can be associated with soil, climate, and plant interactions with the environment. The availability of plant nutrients is a complicated process and it can be classified as a dynamic process. The nutrient efficiency can be linked with the rhizosphere. In the rhizosphere, there are biological, physical, and chemical changes. The physical changes in the rhizosphere can be based on structure, water holding capacity and temperature. The chemical changes in the rhizosphere involve allelopathy, nutrient concentration, nutrient availability, secretion of the compounds from the roots, redox potential, and pH. The field crops can be called agronomic crops and some of the crops can be used as raw materials, livestock feeding and for human consumption.

The crop plants can be classified according to the practices followed in enhancing crop yields. The botanically and agronomically important crops can be classified as dominant crop types and most of the farmers consider these crops as beneficial. The crop plants can be used in medicines, sugar, oil, fiber, forage, and food purposes. Out of these crops, legumes and cereals are important. The botanical classification can be classified as an important system in the crop classification process. The crop plants can be classified into two main groups and it is based on fruits and inflorescence characteristics. The main families are the legume family and grass family.

The grass family is classified as Gramineae and the legume family ps classified as Leguminosae. The grass family was classified as a large family with more than 10000 species (Pegler, 1976). Rye, oats, barley, corn, rice, and wheat were classified as grain crops and they belong to the grass family. Most of the grasses have a fibrous root system. They have panicle, leaves and other systems to survive.

1.2 NUTRIENT UPTAKE

The mineral nutrients are important in the growth and development of plants. These nutrients can be in low concentrations in the soils. Most of the plants have developed nutrient uptake mechanisms that help them to absorb the nutrients as and when required.

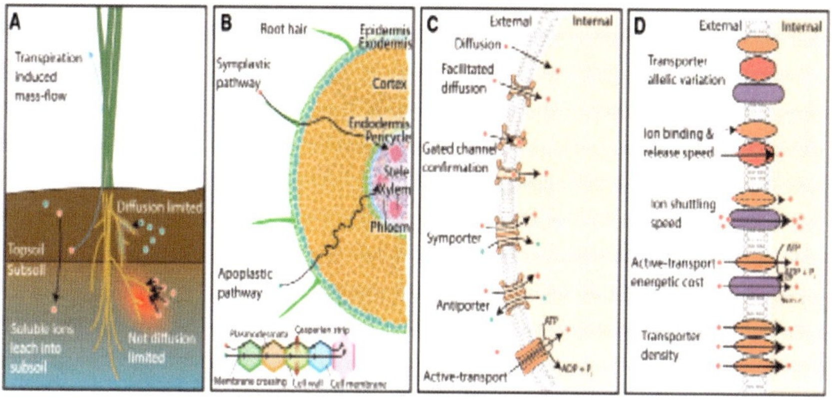

Figure 1.2: Crop nutrient uptake.

Source: http://www.plantphysiol.org/content/182/4/1854

The uptake of the nutrients (Figure 1.2) and transport is regulated by the genes. There are different gene types that may regulate each nutrient. The transporter proteins can be useful in absorbing the nutrients from the soils and their transport via the plasma membrane. In case of nutrient deficiencies, the genes will be activated and there will be the production of mRNA so that the proteins can be produced. These proteins can be called as transporter proteins. There are several metabolic enzymes that are downregulated and, in some cases, they will be upregulated if the plant nutrients are less. Intermediate metabolites, nutrients, growth regulators and amino acids can repress or induce the genes that can encode the transporter proteins. The cells have a plasma membrane and it has several channels and pores. These channels can control the passage of ions and water. The water molecules can be regulated by proteins called as aquaporins or AQPs. These proteins are present in prokaryotes and eukaryotes. The transporter proteins can be classified into 2 classes. They are ion channels and ion transporters. The ion transporters can be called as carriers.

1.3 NUTRIENT USE EFFICIENCY

Climate change and agricultural land use can produce several negative impacts on food productivity. The cropping area must be increased and this will impact the environment negatively. In cereals, the nitrogen use efficiency (Figure 1.3) or NUE was 35% (Raun et al., 1999). Due to advancements in plant breeding techniques, crop yields were increased drastically.

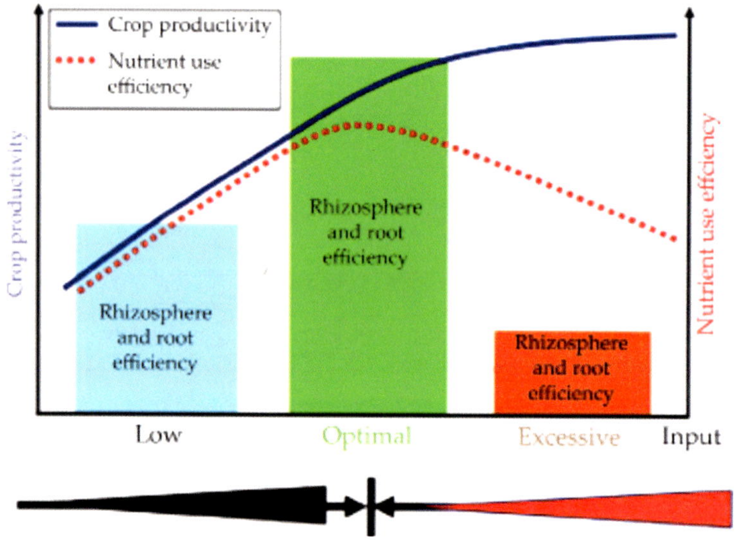

Figure 1.3: Nutrient use efficiency.

Source: https://www.sciencedirect.com/topics/agricultural-and-biological-sciences/nutrient-use-efficiency

The sugarcane yield was enhanced due to plant breeding techniques and wheat production was also increased. The increase in crop yields can be stated as a measure of NUE. The breeding mechanism to produce stress tolerance has improved crop yields. The NUE values can be associated with the input of the fertilizers. The product of utilization efficiencies and uptake can be defined as nutrient use efficiency. An increase in the application of nitrogen can enhance the crop yield and this is possible only if there are no limiting factors and environmental constraints. In several crops, the crop yields will be linked with the photosynthetic process and the products generated. As the crop yields are high, the less important nutrients will be accumulated in the grains. We can observe a lowered concentration of nutrients in the grains and this is because the starch molecules will dilute these nutrient concentrations. The capture or absorption of nutrients can be considered as a root trait. The roots can serve as sinks and they can be sites where the nutrients will be deposited. The uptake of the nutrients can be based on the architecture of the root surface. Some of the nutrients can be present in the soil solution and it is based on the released soil minerals that are soluble. Some of the organic matter can be mineralized and this will contribute to the nutrient uptake mechanisms. The supplement of the nutrients can force the roots to proliferate. The proliferation of the roots can be based on transcription factors. The acquisition of the mineral nutrients can be based on the transmembrane transport. It was observed that the sulfate transporters can help in the migration of selenite molecules. The same transporters can transport molybdate. The storage of the nutrients in the vacuoles can be considered as an important factor. Preferential uptake of the beneficial ions can be important. In some cases, the production of biomass can be taken as a function of nutrient utilization efficiency.

It is always important to ensure that there is balanced nutrition in every phase of the development and growth of the plant tissues to give good yields. Potassium, calcium, nitrogen, and phosphorus can play a vital role in the crops. These elements must be supplied to the crops continuously. It is obvious that the reproduction and growth of the plants are complicated processes. The environmental factors that affect the agricultural ecosystems and natural ecosystems are diverse in nature. Most of the ecosystems are deprived of nutrients. Phosphorous plays a crucial role in regulating the signaling mechanisms of the cells. Potassium plays an important role in the regulation of stomata and this element is responsible for plant tolerance response against stresses. Some of the stresses that plants encounter are because of pests, diseases, drought, and temperatures. The quality of the crops can be affected by the K, N and P nutrient deficiencies in the vegetables, fruits, and forages. The formation of fruits and seeds can be affected by reduced P_i levels. Reduced K levels can affect the shelf life of the vegetables and fruits.

The macronutrient use efficiency can be called as macro-NUE. This is based on agronomic practices, environmental conditions, and plant genotypes. The agronomic practices comprise of irrigation, fertilization methods, arrangement of plants and density of crops. The environmental conditions comprise of nutrient concentrations, alteration in pH, the texture of the soil, temperature, and rainfall. It is guided by two important processes. They are the utilization of the nutrients and acquisition efficiencies. There are about 90 elements that are absorbed by the plants and some of them are classified as essential (Arnon et al., 1939). The essential element must have an important role in the metabolism processes in the plants. They must be specific and should be replaced by any other element. In the absence of this essential element, the plants may not grow normally. The essential nutrients are necessary in the metabolic activities and growth of the plants. These nutrients can be classified as primary macronutrients and secondary macronutrients. Some of them can be called as micronutrients as they are required by the plant in lower quantities.

Potassium, phosphorus, and nitrogen can be classified as primary nutrients. These nutrients can be supplied by the fertilizers to rejuvenate the soils. SulfurSulfur, magnesium, and calcium can be classified as secondary nutrients. They are required by the plant in moderate concentrations. The micronutrients are required by the plants in trace amounts. They are classified into micronutrient cations and micronutrient anions. The micronutrient cations are copper, zinc, manganese, and iron. The micronutrient anions are chlorine, molybdenum, and boron. The mobility of the nutrients can be based on the type of soils. The mobility of the nutrients can determine their availability to plants. In order to improve nutrient absorption, it is essential to know the mobility parameters of the nutrients.

The nutrients can be classified into immobile, less mobile, and mobile. The immobile nutrients can be firmly linked with the soil particles. These nutrients cannot be released easily. Due to this, the immobile nutrients are sparingly available to the plants. Some of the immobile nutrients are $H_2PO^-_4$, HPO^{2-}_4, Cu^{2+}, Zn^{2+}, Mn^{2+} and Fe^{2+} ions. The moderate or less mobile nutrients are soluble and found in the soil solutions. These nutrients can be absorbed by the clay particles. Due to this, they can be dissolved in the soil solution. Some of the less mobile nutrients are MoO^{2-}_4, Mg^{2+}, Ca^{2+}, K^+ and NH^+_4 ions. The mobile nutrients are soluble in water and they can be present in the soil solution. They exhibit high rates of mobility. They are available to the plants. As they are easily available to the plants and have high mobility rates, they are susceptible to leaching. Some of the mobile nutrients are H_2BO_3, Cl^-,

SO_4^{2-} and NO_3^-. The mobile nutrients can translocate to the younger cells from the older cells. This is the reason why the symptoms are dominant in the older leaves than the younger ones. Some of the mobile nutrients are magnesium, potassium, phosphorous and nitrogen. The immobile nutrients cannot get translocated from the younger leaves to the older leaves. Hence, the symptoms will be dominant in the younger leaves. Some of the immobile nutrients are molybdenum, boron, zinc, manganese, iron, sulfur and calcium. Most of the plant body is made up of carbon, hydrogen, and oxygen.

1.4 NUTRIENT DYNAMICS

Most of the agroecosystems are affected by the atmospheric effects and the composition of it. The crops rely on the atmosphere for gaseous exchange and water. The distribution of particles and dust will affect the agroecosystems. (Figure1.4)

The precipitation events can produce soil water and it will be absorbed by the plants. Most of the water that gets impregnated with the soil profile will be absorbed by plants. The precipitation can transport dissolved nutrients, small particles and dust and the concentration of these particles will be based on the geographic location. The dust and deposits in the atmosphere will carry nutrients into the crop fields. Nitrogen can be deposited in the soils from diverse sources. In the agrarian zones of Europe, the nitrogen is derived from the atmosphere. It is rare that plants grow in environmental conditions that are optimal and standardized. As far as crop productivity is concerned, we cannot control all the environmental conditions that enhance crop yields. The physical and chemical fluctuations in the environment can limit the yields. The temperature and moisture variations can comprise the physical factors and nutrient toxicity, deficiency, imbalances, and pH constitute chemical factors. The breeding strategies and genetic manipulations can enhance the uptake of the nutrients and these advanced techniques can also enhance the utilization efficiency. The environmental pollution and soil erosion must be reduced so that crop productivity will be enhanced. The selection techniques are being used to enhance nutrient uptake. In the case of the selection process, Mahon has framed certain criteria (Mahon, 1983).

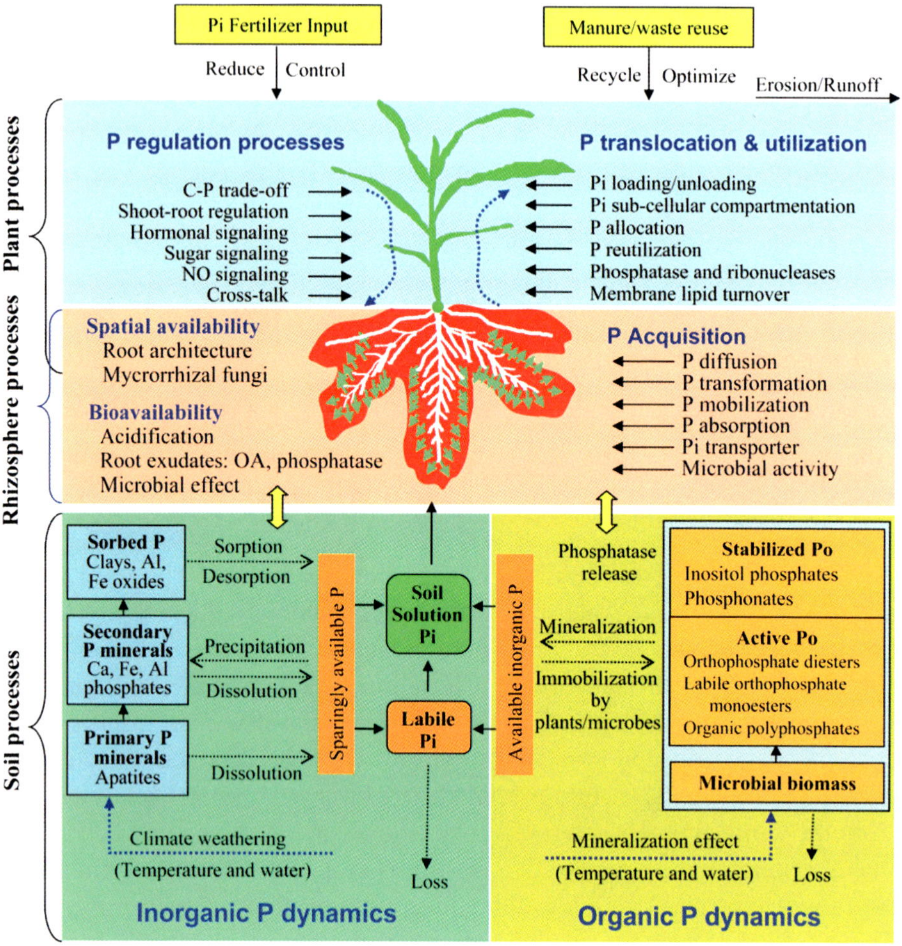

Figure 1.4: Plant nutrient dynamics.

Source: http://www.plantphysiol.org/content/156/3/997

It is important to use the traits that can be used in expressing beneficial characters that can enhance crop yields. It is important to store germplasm in large amounts. It is always important to select improved physiological traits and nutritional traits. It is important to lower the production costs and enhance the stability of the yields. The stability of the genotypic performance must be considered. The genetic variability should be assessed. The genetic diversity must be considered so that we can determine the tolerance of the crops against the toxins. There are several genes that can induce tolerance at various toxicity levels. The plant responses to stress and nutrient uptake mechanisms can be linked. The heritability process can explain the genetic portion of the genome and this should be supported by the variance obtained in the statistical studies. The breeding conditions can be determined by the heritability rates. The genes that are classified as recessive can interrupt the important processes and due to this the enzymatic activity will be lost and so the phenotypes produced will be useless. The reduced nutrient adaptability can be explained by the failure of roots to absorb a specific nutrient. The management of soil fertility can be important in the production of the crops.

The response of the plants towards essential nutrients must be studied in detail. There are three phases that determine the response phases. They are the deficiency phase, tolerance, and toxicity phase. There are several external factors and internal factors that are important in the breeding programs. The internal factors can be detoxification, assimilation, translocation, and absorption. The external factors include moisture stress, soil acidity and root exudates. In case of high nutrient availability, the crops will go for rapid nutrient absorption and so there will be rapid growth. The nutrient and tissue ratios will be normal. In the case of high nutrients availability, the wild plants will increase their nutrient absorption. There will be a slow rate of absorption. The nutrient to tissue ratio will be high. Due to this, there will be nutrient toxicity. In the case of low nutrient availability, the crops exhibit reduced nutrient absorption. We can observe continuous growth. The nutrient to tissue ratio will be low. Nutrient deficiencies can be observed in this case. In reduced availability of the nutrients, the wild plants exhibit slow rates of nutrient absorption. The rate of growth will be slow. The nutrient to tissue ratio will be normal. Due to the increased metabolic efficiency, the root volume will be increased. The cation-exchange capacity or CEC and exchange absorption of the roots can be important.

The nutrition of the plants can be controlled by the genes and the genetic control can be simple in the case of micronutrient absorption and complex in the case of macronutrients. In the growth cycle of the plants, the rate of micronutrient absorption will be slow. The soil organic matter can be determined and enhanced by an increased input of carbon and loss of carbon molecules. Due to increased carbon inputs, there will be increased biomass. This is associated with the application of legume-grass mixture, irrigation, fertilizer, selection of species and the productivity of forage.

Figure 1.5: Plant biotic stresses.

Source: https://orbitbiotech.com/temperature-stress-plants-high-temperature-stress/

There are many biotic (Figure 1.5) and abiotic stresses (Figure 1.6) that affects the growth of the plants. These stresses can regulate crop yields. The pollution, imbalance in nutrients supply, and starvation can be classified as abiotic stresses. Some of the climatic factors like low temperatures, heat waves, flooding, and drought are considered as abiotic stresses. The climate factors can control the plant nutrition uptake. There will be one or more stresses that can affect the growth of crops. The knowledge of stress periods should be essential to monitor the impacts of stress. It is important to maintain a balance between the micronutrients and macronutrients so that the plants may not get affected by abiotic and biotic stresses. The stresses caused by heavy metals and drought can be managed by silicon. There is a link between photosynthetic efficiency and the supply of magnesium). The transport of the nutrients can be observed in the reproductive parts of the plant and vegetative parts (Haruna et al., 2020).

Figure 1.6: Plant abiotic stresses.

Source: https://orbitbiotech.com/temperature-stress-plants-high-temperature-stress/

The mobility of the nutrients in the phloem tissues can be important for plant growth. The redistribution processes can be important in the case of toxicity, hyperaccumulation, and heavy metals (Hoduck et al., 2018). The deficiency of zinc in plant products has become a serious concern these days. There were considerable changes in the growth patterns of the maize crops due to reduced absorption of K, P and N. The stresses can induce a structural response from the plants, and due to this, they will become sensitive and prone to diseases.

Many materials can help the plants to absorb the nutrients. Some of the nutrients can be natural and some materials are synthetic. We can use the recycled materials as nutrients and in some cases, we can use microbial inoculants. Most of the nutrients can be absorbed by the plants easily. We can classify the sources of nutrition into biological, mineral, and organic. Most of the organic nutrients can be classified as manures and organic fertilizers. Some of the residues that are produced from animal and plant products can be used as nutrients and the plants use nutrient recycling mechanisms. Most of the nutrients can be derived from mineral fertilizers. The fertilizers that can supply nutrients can be classified as straight fertilizers and complex fertilizers. The straight fertilizers can comprise one of the nutrients like K, P, and N. These fertilizers can be called as single-nutrient fertilizers.

The complex fertilizers comprises of two nutrients. Most of these nutrients are produced from raw materials. They are produced in the form of solids and are granulated. They are classified as three nutrient fertilizers like NPK and two nutrient fertilizers like NP. Most of the nutrients can be described in their elemental forms. Some of the industries produce oxides of K and P. Some secondary crop products comprise nutrients of low grade. These nutrients can enhance soil fertility. The composting process can enhance the quality of the nutrients. The oil cakes can have a larger nutrient content than the raw crop residues. Most of the crop residues are the leftovers that are present in the croplands after the harvest. These residues include husks, trash, stubble, stalk, and straw. They are having many uses after composting and these materials can be used for enriching the plants. The composition of straw is higher than the actual product and so we can use it easily. Most of the crop residue can comprise of high nutrient concentrations.

In the case of sugar cane and sugar beet crops, most of the leaves are left in the crop fields as they serve no immediate purpose and these leaves can be used as a good source of nutrients for the same crops and the fertility of the soil will be enhanced. Oil cakes can be classified as an important crop residue that can be used in many crops. The oil cakes can be called as residues that remain after the oil is taken out from the seeds. Most of the -edible oil cakes can be used as manure and if they are edible, we can use them in the form of cattle feed. The proportion of phosphorus and nitrogen are high in oil cakes. The fresh green plant materials can be used in the form of green manures. Most of the legumes are grown for this purpose so that the fertility of the soils can be restored. The leftover from the legume crops can be used as green manures. The green manures obtained from the legume crops have high proportions of nitrogen and are used in integrated nutrient management. (Figure 1.7)

The green manure can be prepared as a part of the crop itself or it can be prepared elsewhere and brought into the crop fields. Some plants cannot be used as green manures. Green manures are added to crops like sugar cane, potato, and rice. We can use legumes as both short-term crops and long-term crops. These crops can be grown before taking out the pods or after that.

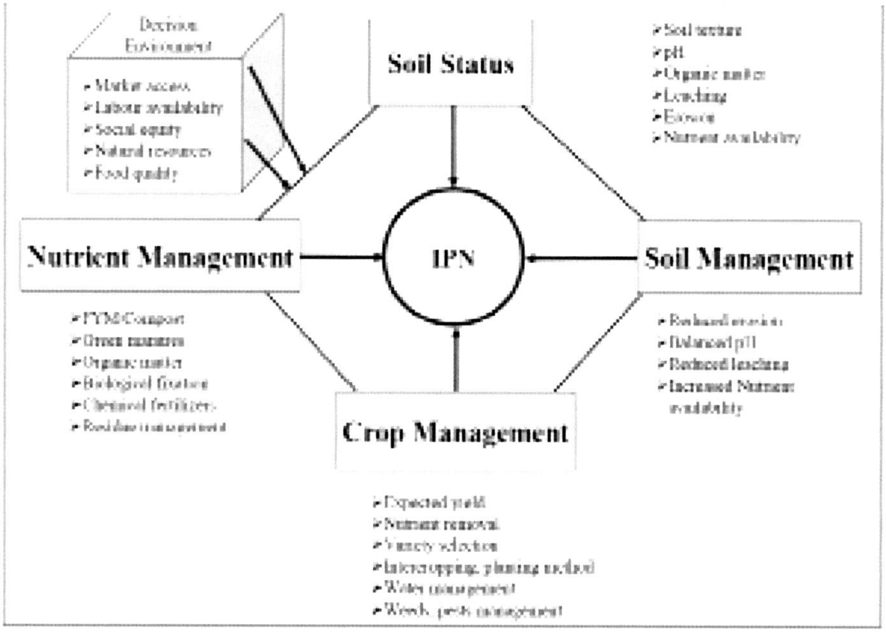

Figure 1.7: Integrated nutrient management.

Source: http://www.fao.org/3/AG120E10.htm

The farmyard manure or FYM (Figure 1.9) can be defined as organic manure that are decomposed naturally. It can be prepared from the urine and dung of the animals. We can use litter as bedding material. Whilst storage, organic manure can be decomposed by the fermentation process. Due to the partial decomposition of these substances, there will be the production of the humic compounds that are classified as useful. The FYM can be used in crops like pearl millet, maize, and rice. Animal slurry is mostly used in developed countries. This is true because of intensive agriculture with the use of less labor. In most of the areas, the use of slurry has become a dominant practice.

Biogas plant slurry is being used in several countries to supplement the crop nutrients. We can use organic wastes in the production of biogas. In developing countries, there are several biogas plants. In these biogas plants, we can use cattle dung as the raw material. Balanced C: N ratio was observed when elephant dung was mixed with human waste and this enhanced the biogas production (Karki et al., 2013).

Figure 1.8 Green manure.

Source: http://agritech.tnau.ac.in/org_farm/orgfarm_green%20manure.html

Several organic wastes can be applied to the soils directly. Due to the addition of these wastes, the soil components will be enriched with the nutrients and are available for plants. The fertilizers and organic manure that can be produced due to the decomposition of industrial wastes, human and animal wastes are called as compost. Rural compost can be produced in the farms being managed in rural areas. The sewage sludge, city garbage, and industrial wastes can be subjected to the composting process to produce urban compost. Vermicompost can be produced due to the action of earthworms on the compost. The vermicompost comprises of hormones, enzymes, organic matter, and plant nutrients. We can observe actinomycetes and microbes that are beneficial. The preparation of the compost can be through different microbes. The microbes can act on human wastes, kitchen wastes, sawdust, bagasse, weeds, crop residues, leaves, roots, and other crop residues. Most of the biodegradable organic wastes can be subjected to microbe-assisted degradation. The concentration of the nutrients in the compost can be based on the waste quantities.

Figure 1.9: Farmyard manure.

Source: https://www.agrifarming.in/farmyard-manure-preparation-methods-advantages

Biofertilizers (Figure 1.10) can be defined as materials that comprise living microbes or resting microbes like algae, actinomycetes, fungi, and bacteria. The biofertilizers enhances the absorption of the nitrogen molecules and their supply to the crops as growth-promoting substances. The biofertilizers are called as microbial cultures and bioinoculants. We can classify biofertilizers into 4 major classes. They are plant growth-promoting rhizobacteria or PGPR, Composting accelerators, phosphate solubilizing biofertilizers and nitrogen-fixing biofertilizers. PGPR comprises of Pseudomonas bacteria in excess. These biofertilizers increases plant growth and they are not involved in the supply of plant nutrients. The composting accelerators can be classified into cellulolytic and lignolytic accelerators. Trichoderma comes under a cellulolytic accelerator and Humicola can be classified as a lignolytic accelerator. The phosphate solubilizing biofertilizers can be classified as PSB or phosphate solubilizing bacteria and PSMs or phosphate solubilizing microbes. Mycorrhiza like VAM can mobilize the nutrients. The nitrogen-fixing biofertilizers are *Acetobacter*, *Clostridium*, *Azospirillum*, *Azotobacter* and *Rhizobium* species. The Azolla species and cyanobacteria can be used in fixing nitrogen. Nitrogen-fixing bacteria or microbes can be classified as important species in describing the biofertilizers and phosphate solubilizing fertilizers can be treated as a second priority.

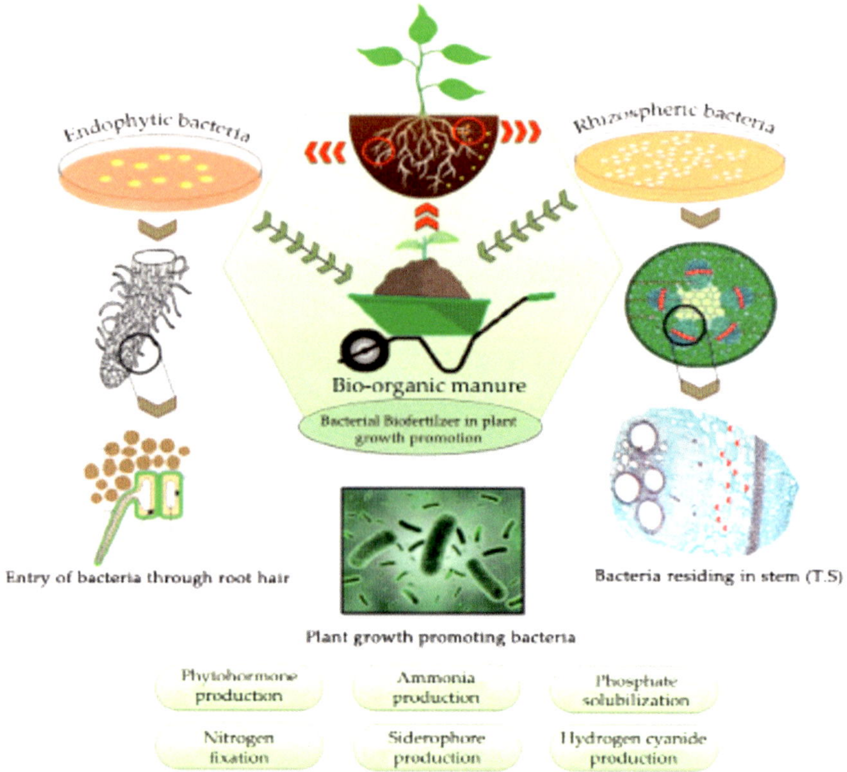

Figure 1.10: Biofertilizers.

Source: https://www.sciencedirect.com/topics/agricultural-and-biological-sciences/biofertilizers

1.5 ENVIRONMENTAL ISSUES

Nutrient management affect the environment in many ways. The improvement of fertility and its depletion can be considered as an improvement and degradation of the environment. The depletion of soil fertility is high than its improvement. The nutrients that are being added through composts, manures, and fertilizers can impose both negative and positive effects on the environment. The nutrients can be absorbed by the plants and some of the nutrients are immobilized and may not be available for plants. In some cases, the nutrients can get volatilized into the atmosphere and some of them can get lost due to leaching and erosion. Some of the nitrogen compounds can be released into the atmosphere denitrification process. The use of plant nutrients enhance crop yields and due to this, the imbalance created can be nullified. Some of the nutrient deficiencies in the soils can be adjusted. Some of the nutrients that are extracted from soils whilst the harvest season can be balanced. The plant nutrients should be used judiciously so that the environmental pollution load will be reduced. Due to the immobilization of the excess nutrients in the soils, environmental pollution will get increased. The fsulfates and nitrates may not firmly bind to the soil particles and so they can get leached with water and this will produce several environmental problems.

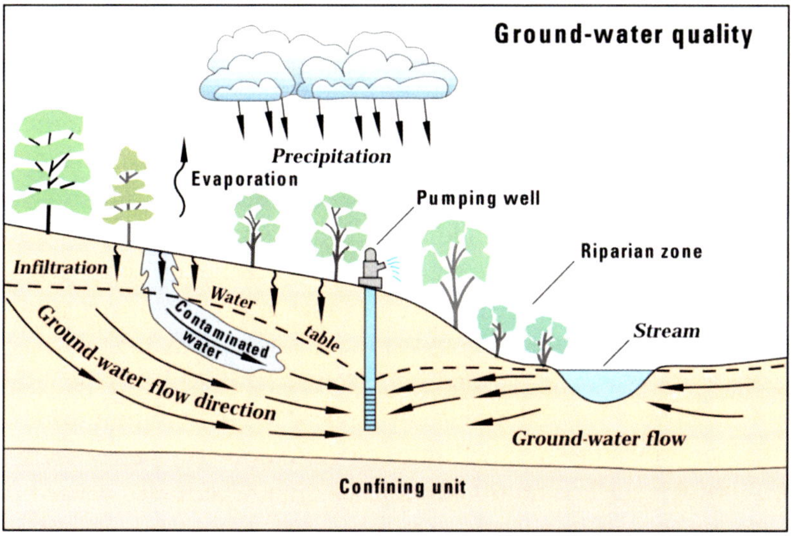

Figure 1.11: Groundwater contamination.

Source: https://www.usgs.gov/special-topic/water-science-school/science/pesticides-groundwater?qt-science_center_objects=0#qt-science_center_objects

Groundwater contamination (Figure 1.11) can be considered as an important problem if excess nutrients are used. Due to leaching, boric acid, sulfate, chloride, and nitrates can get accumulated in the environment. It is important to use fertilizers according to the need. It is important to ensure optimal loading rates. The nitrogen efficiency must be enhanced using sustainable practices. It is important to follow integrated water management and integrated nitrogen management. Eutrophication is another problem with these excess nutrients. The nutrients will be eroded due to groundwater discharge and surface runoff. It is important to regulate and minimize soil erosion. We need to minimize the runoff by growing crops and practicing water harvesting techniques. Irrigation measures should be regulated. Extreme nutrient supply can lead to methemoglobinaemia in humans and cattle. This occurs due to an overconsumption of nitrates via food and drinking water. This condition can be avoided or minimized by enhancing the quality of water and by minimizing the losses due to leaching.

The deposition of ammonia and acid rain can be classified as important problems caused by the application of excess nutrients. The moisture present in the atmosphere can react with the oxides of N and there will be the formation of nitric acid. There will be increased emission of sulfur and ammonia will be volatilized in large amounts. This problem can be reduced by lowering the denitrification process. We need to alter the soil pH so that the plants can survive the losses incurred due to depletion of the nutrients. Global warming and depletion of ozone can occur due to the excess use of nutrients. This can be caused due to the emission of nitrous oxide from the soil particles. To reduce these problems, we can use chemicals that can inhibit the nitrification process. The addition of nitrogen can cause a significant change in crop yields and there are several environmental concerns that are to be addressed. The nitrogen molecules that are immobilized in the soil particles and unavailable to the crops can be removed from the soils in various ways. The nitrates can be leached into groundwater. Most of the ammonia molecules can be released into the atmosphere in the form of nitrous oxide. The nitrogen molecules applied through fertilizers can be removed from the soil through surface runoff and soil erosion. The losses caused due to volatilization, runoff and leaching can be high. Nitrogen consumption all over the world was about 84.7 million tonnes in 2002 (FAO, 2003). The organic matter present in the oils can be eroded into the environment. It was estimated that at least half of the N minerals are lost to the environment each year. Nitrates are most susceptible to erosion. Some factors allows nitrate leaching into the water. This can be due to unregulated floods, higher water table, excess rainfall cracked soils, and high nitrate concentrations. The leaching of N has several negative effects.

The main problem with the nitrates is that whist leaching, the chlorides, sulfates, and nitrates will take most of the cations with them. The cations that are most susceptible are potassium, magnesium, and calcium ions. Due to the loss of these cations, there will be an increase in the acidity of the soils. Due to the volatilization of ammonia from vegetation and soils, there was dumping of about 21 million tonnes of N into the atmosphere (Smil, 1999). The volatilization of ammonia can occur due to several factors. Some of the factors are shallow penetration, high fertilization rates, high wind velocity, high soil temperatures, liquid fertilizer, high soil pH, and soils that have high calcium carbonates. The ammonia molecules that are volatilized can get back to the soils via snow or rain. This can be possible due to the N cycle. The animal manure that is in liquid form can lose high N concentrations. The nitrous oxides and nitrogen monoxides can be produced by the microbes that are present in the soils. This is possible if the oxygen supply is less. This process can be called as denitrification.

Phosphate can be present in the soils in both inorganic and organic forms. They have differential rates of mobility and solubility. Applying nutrients at a normal specified range can get accumulated in the soils as time progresses. The phosphorus molecules that are being accumulated can be useful for the plants.

Chapter 2

Essential Plant Nutrients

CONTENTS

2.1 Introduction 26

2.2 Nitrogen 30

2.3 Phosphorus 33

2.4 Potassium 35

2.5 Calcium 36

2.6 Magnesium 37

2.7 Sulfur 38

2.8 Zinc 40

2.9 Copper 42

2.10 Iron 43

2.11 Manganese 44

2.12 Nickel 46

2.13 Chlorine 46

2.14 Molybdenum 47

2.15 Boron 48

2.1 INTRODUCTION

The plants need many nutrients in their growth and metabolism. The source of nutrients can be from the soils and in the case of lab experiment like in micropropagation via tissue culture, we can use the rooting medium that can be prepared artificially. The root hair cells are considered as important in the absorption of nutrients. These cells have a plasma membrane through which the nutrients enter the plants and participate in the metabolism. The absorption of the plant nutrients can be controlled by many genes and each nutrient uptake will need a specific gene set. Several transporter proteins are present in the plasma membrane. These proteins allows the nutrients to enter the cytoplasm. Many enzymes are needed for the regulation of the nutrients in the cytoplasm. There are intermediate metabolites, growth regulators, and amino acids with nutrients that can be needed for the activation of the genes that encode for the expression of transporter proteins.

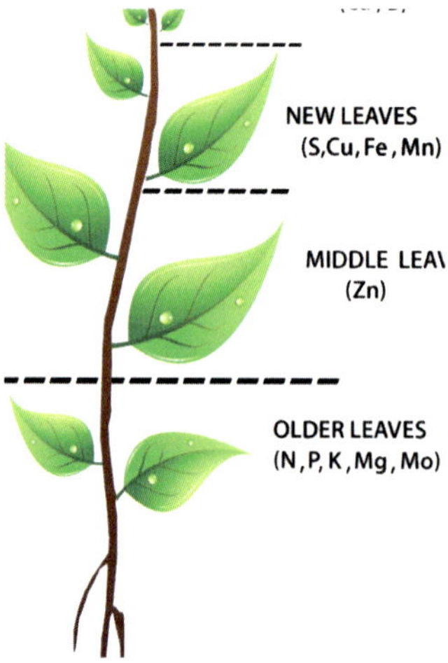

Figure 2.1: Essential plant nutrients.

Source: https://blog.agrihomegh.com/essential-plant-nutrients/

Most of the nutrients (Figure 2.1) are absorbed by the plants in their ionic forms. These elements are present in the soils and growth mediums. The nutrients that are needed for the plants for metabolism and growth are called as essential nutrients. The essential nutrients are those nutrients without which the plants express deficiency symptoms. There are about 18 essential elements that were recorded to date. They are carbon, hydrogen, oxygen, nitrogen, phosphorous, potassium, calcium, magnesium, sulfur, iron, zinc, copper, manganese, boron, chlorine, molybdenum, cobalt, and nickel. These essential nutrients can be classified into macronutrients and micronutrients. The macronutrients are needed by the plants in large proportions. They are classified as structural elements, primary nutrients, and secondary nutrients. The structural elements are oxygen, hydrogen, and carbon. The primary nutrients are potassium, phosphorous, and nitrogen. Magnesium, calcium, and sulfur are classified as secondary nutrients. Some of the micronutrients that are needed for plants are nickel, cobalt, chloride, molybdenum, boron, copper, manganese, iron, and zinc.

The sodium ions are important in the case of halophytes. The salts get accumulated in the vacuoles so that the growth and turgor will be managed. Sodium is essential in C_4 plants. The carbon dioxide can be concentrated by sodium ions. Due to the presence of sodium, there will be glossy and green leaves. Due to increased sodium concentrations, the taste and texture of the crops will be increased. The silicon molecules helps in strengthening the stems. Due to the presence of silicon, the plants become tolerant to abiotic stress and biotic stress. Cobalt helps crops in the process of nitrogen fixation. It helps diazotrophs and root nodule bacteria. Selenium deposits in crop plants and their consumption by humans reduces the risk of heart disease and cancer. Some non-essential nutrients plays an important role in plant metabolisms. Most of the plants grow in ionic environments that are considered as hostile. This condition is seen in soils that are rich in minerals. Some of the non-essential nutrients that are in their ionic form are zinc, strontium, nickel, manganese, ferrous, ferric, cobalt, copper, cadmium, aluminum, and gold. The ionic forms can be considered as toxic to the plants if they are present in high concentrations. Some of the elements like uranium, lead, cadmium, mercury, chromium, and others have no specific function in the metabolism. Some of these elements can be anthropogenic and some can naturally originate according to local geological structures. The soils that were extracted from the waste dumps and mining sites can be toxic and may destroy the plants. If the food crops are grown in these soils, the toxic elements can enter the food chain and so the animals and humans will get affected. Some plants have developed certain mechanisms to detoxify the chemicals. These plants can shift the metals absorbed by the plants into vacuoles. The root cells can scavenge the metal ions and there will be chelation in the cytosols. Many transport systems can control the transport of metal ions. Radioactive nuclides can be considered as a portion of the environment that can be derived from the cosmic rays and the earth's crust. The cesium molecules can be accumulated in the soils and they can stay intact for many years and thus the soils are not available for crops. The plants and animals can survive in the areas that have high rates of radiation only if human interference is at a minimum. Arabidopsis plant has developed to the mutagens. The human environment produced 85% of the radiation dose in 2015 (World Nuclear Association, 2015). It was observed that the arable soils are limited by plant nutrients. The concentration of the nutrients in the soils can be based on local climatic conditions, soil characteristics and removal of nutrients caused due to irregular drainage and irrigation practices. The farming practices can also alter the nutrient profile and it is also based on the soil types. The demand for nutrients will get changed according to the growth phases. The growth stages of the plants demand more nutrients that the soils may not supply. The plants have developed several mechanisms so that they can grow in nutrient-limiting conditions. Most of the plants absorb mineral nutrients for their growth and metabolism. In plants, many types of genes can be induced in the case of sufficiency and deficiency of nutrients. This can be done by nutrients, metabolites, plant growth regulators, and amino acids. These genes can regulate transporter proteins.

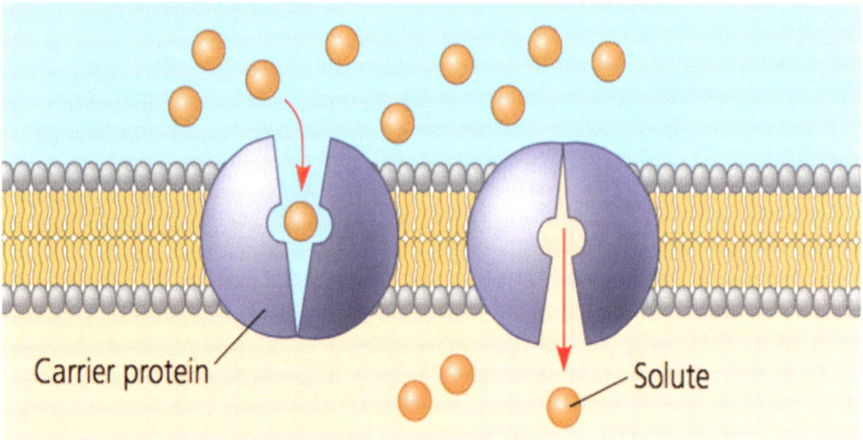

Figure 2.2: Transporter proteins.

Source: https://apbiologyctd.wordpress.com/chemistry-and-cells/membrane-structure-and-function/

The transporter proteins (Figure 2.2) can be of two types. They are channel proteins and transporters. We can observe many pores in the plasma membrane and due to these pores, specific molecules and nutrients can enter the cells. The channel proteins have alpha helices that are present across membranes and they are classified as large molecules. Gating conformations like closed conformations and open conformations can be observed. Some factors can affect the gating process. They are membrane stretch, binding of molecules, and fluctuations in the membrane potential.

Most of the transporter proteins can be classified as vectoral membranes. These proteins help in the transport of specific ions across the membrane and this can be aided by the conformational fluctuations. The primary proteins like potassium, phosphorus, and nitrogen are needed in large quantities for the growth and metabolism of the plants. The arable soils may not have the nutrient reserves and so these nutrients must be replenished from time to time. The nitrogen concentration can be constant in the cytoplasm. The ammonia molecules are toxic to the plants and they can accumulate in the plant cells. Nitrogen is present in the enzymes, chlorophyll, pyrimidine and purine rings, and amino acids. Most of the nitrogenous compounds participate in the growth and metabolism of the plants. The leaves can turn dark green if the rate of photosynthesis is enhanced. If N is less, then the chloroplast will be reduced and the entire photosynthesis process will be disturbed. Due to this, the yellowing of leaves occurs. Genes control the uptake and redistribution along with the transport of nitrogenous compounds or ions like nitrates and ammonium. The water molecules can be absorbed by the cells via water channels. The proteins that can help in this process can be called as aquaporins. They are large proteins and are present in prokaryotes and eukaryotes. These proteins regulate the transport of water via channels that are generally narrow and are present in the plasma membranes. The molecules must be in the correct orientation so that the water molecules can pass through the plasma membranes without any problem. The hydrogen and hydronium ions can be rejected by the channels. The molecules like urea and glycerol can enter the channels. Most of the glycerol molecules can enter the channel in the form of a single file. Potassium, Phosphorous, and nitrogen are needed by the plants in large proportions. The nitrogen use efficiency can be improved by the biotechnology approach. Through biotechnological applications, the genes can be manipulated in the translocation of nitrogen and its assimilation. Phosphorus is an important component of compounds that have high energy. Some of these molecules are ATP, CTP, UTP and GTP, creatinine phosphate, and phosphoenolpyruvate. These molecules can supply energy so that endergonic reactions can occur without any problem. Most of the plants can react or respond to phosphorus deficiency. This response will be in the form of genetic and metabolic effects. In some cases, there will be morphological adaptations. The deficiency of phosphates can induce several genes so that the plants can get these molecules from external sources. The roots can secrete many organic compounds and this occurs in normal conditions. Some of these secretions can be proteins, phenolics, growth hormones, amino acids, and sugars.

2.2 NITROGEN

Nitrogen is classified as an important nutrient in the production of crops all over the world. It is required by many crops in higher amounts. The farming systems that depend on the soil for nitrogen cannot support agriculture for long periods. Most of the nitrogen supplements needed for crop growth are supplied in the form of fertilizers. It was estimated that about 80 million metric tonnes of nitrogen supplemented fertilizers were produced (Esteller et al., 2009). Nitrogen fertilizers have a positive effect on human health as the nutritional quality is increased in the crops. These foods are being used all over the world. Several microbes can immobilize the nitrogen molecules and it is common in the case of undecomposed plant residues. Most of the nitrogen molecules are lost in the form of ammonia molecules from the canopy.

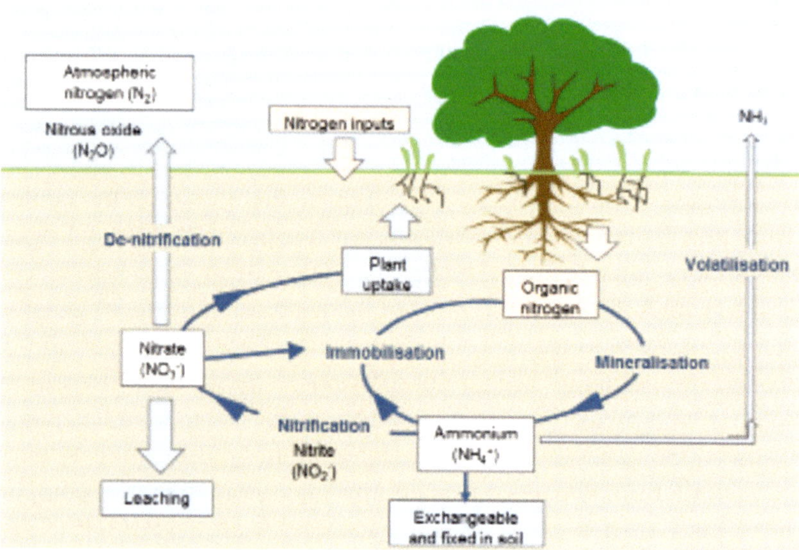

Figure 2.3: Nitrogen cycle.

Source: https://kids.frontiersin.org/article/10.3389/frym.2019.00041

Intensive cropping patterns have resulted in decreased organic matter. The reduction in organic matter can lead to lowered crop production. Sowing the crop cultivars that can yield more must be used by the farmers so that the N-deficiency can be compensated as these cultivars can adapt to the nitrogen limiting conditions. There was a considerable increase in nitrogen losses. Due to increased N losses, the expenditure incurred in the production of crops will be increased and these techniques will harm the environment. In legumes, on the application of N-based fertilizers, we can observe an increased number of pods and in the case of cereals, there was an increase in heads and panicles.

The nitrogen cycle (Figure 2.3) can be classified as a complex and dynamic system that affects the plants, soil, and climate of a region. The nitrogen molecules can be supplied to soils through biological nitrogen fixation, organic manures, and chemical fertilizers. Some of the factors that affect the depletion of the nitrogen sources are surface runoff, volatilization, denitrification, and leaching. The surface runoff is regulated by the precipitation patterns of a region. There will be an immobilization of nitrogen and this can be considered as a transformation process that involves the conversion of the N compounds that are inorganic into ammonia, nitrites, and nitrates. In this case, the inorganic nitrogen is transformed into organic forms. Most of the organisms that survive in the soils can fix inorganic nitrogen and transform them into organic forms. These organic N compounds can be present in the tissues and cells and in some cases as biomass. The nitrogen content in the mineral soils at the surface can be about 5 grams per kg (Brady et al., 2011). Most of the nitrogen molecules are present in the form of organic matter. Most of the nitrogen molecules that are present in the organic forms cannot be used by the crop plants instantaneously. The organic N must be converted into nitrates so that the plants can use them.

The organic N forms can be converted into ammonium ions and nitrates. If the organic N is converted into NH^+_4 then we call it as ammonification. In some cases, ammonium compounds can be oxidized to form NO^-_3 and this process is called as nitrification. In the nitrification process, Nitrosomonas converts ammonia into nitrites and Nitrobacter converts the nitrites into nitrates. The nitrification process can occur at faster rates if the humidity and temperature of the soils are at optimal ranges. The aeration process enhances the rate of nitrification. Aeration in the soil can be high if the cultivation and plowing are done in frequent intervals. The acidification of the soils is possible only if there are H^+ ions in the soils. These H^+ ions can be supplied by the nitrification process. The nitrate ions are negativelycharged molecules. Due to irrigation and rainfall, these molecules are prone to leaching. The leaching of N can be based on the application modes of the fertilizers, soil type, and the N source. The nitrates can associate with potassium, magnesium, and calcium, and whilst leaching these compounds can be leached from the soil. When these bases are leached, there will be acidification of the soils. The denitrification process can be considered as a major loss of nitrogen from the soils. Most of the anaerobic bacteria are involved in the denitrification process. These bacteria are populated in the topsoil that can extend to 30 cm deep (Parkin et al., 1989). The denitrification process can be affected by water, aeration, nitrate concentration, organic carbon, temperature, and the pH of the soils.

The denitrification process is common in the grasslands. Every year, there is an increase of 0.3% increase of N_2O in the atmosphere (Seiler, 1988). Most of the N emissions are derived from the soils and this can be due to denitrification and nitrification. The soils can be considered as an important source from which N_2O will be accumulated. The volatilization of ammonia can be considered as a key process from which the loss of N is higher. This is predominant in alkaline soils. Most of the volatilization of ammonia will be in the rice fields as they will be flooded. This is the main reason behind low nutrient use efficiency in rice crops. The volatilization of ammonia can be affected or regulated by several factors. Some of the factors are wind speed, water depth, temperature, pH, and NH_4-N concentration. The loss of NH_3 can be based on the growth stage of the plants and the status of N. It was observed that the volatilization of NH_3 can be higher in the reproductive stages. Ammonia fixation can be classified as an important portion of the soil and plant systems. Most of the ammonium fixation can be in anhydrous form. Due to an increase in the pH levels, there will be increased NH_4^+ fixation. Most of the nitrogen molecules can be stored in the seeds of legumes and cereals and these molecules may not be recycled. The crops that are nitrogen deficient can exhibit slow growth rates. Due to reduced nitrogen levels, there will be less photosynthetic activity and lowered radiation use efficiency. There is a link between the nitrogen concentration in the plants and the light saturation rates in the photosynthesis process.

The crop nutrient efficiency can be studied in the forms of agronomic efficiency or AE, physiological efficiency, or PE, an agrophysiological efficiency or APE, apparent recovery efficiency or ARE, and utilization efficiency or UE (Fageria et al., 2011). The uptake of the nutrients can be determined by analyzing the tissues. The accumulation of the nutrients and their accumulation can be considered as important whilst this analysis. The product obtained for each unit of a nutrient supplied can be considered as agronomic efficiency of that nutrient. The biological yield for every unit of nutrient supplied can be called as physiological efficiency.

2.3 PHOSPHORUS

In the case of intensified agriculture methods in modern agriculture, several opportunities can bring about maximum yields. Nutrient deficiency is considered as an important problem in agriculture and it should be addressed with extreme care as it can affect humans directly and the economies will be disturbed. Phosphorus can affect many agricultural ecosystems and natural systems (Vries et al., 2013). This element is considered as an important for animals and plants. Phosphate-based fertilizers affects food production worldwide.

Figure 2.4: Phosphorous deficiency.

Source: https://gardener.fandom.com/wiki/Phosphorus_deficiency

Phosphorous deficiency (Figure 2.4) in crop plants can be affected by several factors. Some of the factors are soil erosion, low fixation rates and low spraying rates. This deficiency is common in developing countries. Phosphorus deficiency can lead to an increase in weeds, diseases, and insects. These agents can damage the crops and hence the crop yields will be reduced drastically. The acidity in the soils can be considered as a fundamental cause of phosphorus deficiency. The phosphorus molecules can be supplied to the soils and these molecules can get attached to the soil particles. There will be no immediate use of the adhered phosphate molecules to the crops. The use efficiency of phosphorus can be important in enhancing crop yields. Apatite rock with a high percentage of phosphorus and to be used as fertilizers can be considered as phosphate rock or PR (Cathcart, 2015). If P leaves the soils due to erosion or other factors, it is hard to get back the P molecules from the external resources. A limited proportion of phosphorus is present in the soils and it is not readily available to the crop plants. The inorganic P forms can be linked with aluminum, iron, and calcium. These molecules are mostly available in acidic soils. The P molecules should be mineralized before absorption. This process can be assisted by the microbes.

The processes like solubilization and immobilization can be considered as important factors in determining the role of P in plants and soils. Fixation or immobilization is nothing but the strong precipitation and adsorption of the phosphate ions on the iron hydroxides and aluminum hydroxides. The immobilization of the P molecules can be high in soils that have aluminum hydroxides and iron hydroxides. In some cases, allophane can assist the immobilization of the P molecules. Due to the complex forming nature of the P molecules, these molecules are insoluble and may not be available for plants readily. The main problem with the P molecules is not associated with its acquisition but increasing the availability of the P molecules for the crop plants. The precipitation mechanism can be possible if the P levels are high and adsorption reactions or mechanism occurs only if there is a low P concentration.

The soils that are having high clay content can support high levels of P molecules. In the case of Ultisols and Oxisols, the immobilization of the P molecules is high. P can be used by crop plants for both growth and reproduction. It can be used by the plants along with K and N. There was an increase in the root growth on the application of P fertilizers on the legume plants (Baligar et al., 2010). P can enhance the plant strength and so lodging will be prevented. The fruiting and seeding process in the plants can be regulated by the P molecules. This element is important in the meristematic tissues where the growth activity is higher. The germination process in the crop plants can be controlled by the P molecules as they are classified as important molecules in the Phytin component. The P deficiency can lead to reduced seed viability and a lowered number of seeds. This can be due to an energy shortage in the P deficient conditions. There will be reduced leaf expansion, reduced number of leaves, and shrunk leaves if P molecules are less. There was an increased root-shoot ratio in lower P conditions. The shoot branching will be lowered. It was observed that the assimilation and fixation of carbon were severely affected due to P deficiency. The response of the crop to P nutrients can be considered as an important method to evaluate the nutrient use efficiency. There will be dark green leaves, thin stems, and stunting in the plants. There will be a decrease in the photosynthetic efficiency of the leaves. There will be an accumulation of anthocyanins in the leaves and they appear as purple and red.

2.4 POTASSIUM

Potassium (Figure 2.5) can be considered as an important element in crop systems. Its presence in nature is higher than that of soils. Most of the K forms that are needed for the crop plants are not present in the soils. The K deficiency can be considered as an annual problem. This is due to crop rotation and most of the K is lost due to mining. Most of the K compounds are stored in the straw. The straw will be removed and it will be used as fuel and fodder and this will result in loss of K ions in the soils. The supply of K ions can be important to reduce environmental pollution and enhance crop productivity. The production costs will also come down. Potassium salts were first used in 1861 in the form of fertilizers (Sheldrick, 2015). The main sources of K are chemical fertilizers, manures, and crop residues.

Figure 2.5: Potassium deficiency on leaves.

Source: https://www.growweedeasy.com/cannabis-plant-problems/potassium-deficiency

Most of the K ions are eliminated by runoff, erosion, and leaching. K can migrate from the roots to all plant parts through diffusion. This process is based on the concentration gradient, diffusion coefficient, temperature, diffusion path, and soil water composition (Bertsch et al., 2015). The potassium molecules in the soils can be present in the crystal structures of feldspar and mica. The uptake of K by the plants can be considered as a dynamic process. Potassium can enhance the growth of roots. It can also improve nutrient uptake and enhance water absorption. K ions can eliminate lodging and increase cellulose. This molecule becomes a part of about 60 enzymes that are essential for the growth of crop plants. It increases resistance against fungal infestations. It also speeds up the healing process. It affects the transport of iron in dicots and monocots. The opening and closing of stomata can be regulated by K^+ ions. It can eliminate acids that generally occur in metabolic reactions. The nitrogen fixation can be reduced in the legumes due to K deficiency. Chlorophyll molecules will be damaged due to reduced K concentrations. The K levels can be enhanced in the soils only if the fertilizers are sprayed and other inputs are supplied to the soils. The retention capacity is very less. This can be due to a reduced cation exchange capacity or CEC. There will be depletion of the K ions just after cropping seasons. It is important to know the K dynamics in the plants and soils.

2.5 CALCIUM

Calcium (Figure 2.6) can exist as a divalent cation and it is important in the growth and development of plants. The quality and yield of crops can be reduced due to lowered calcium bioavailability. Most of the calcium ions are present in alkaline soils and neutral soils. The efficiency of calcium can be observed in the ultisols and oxisols. They are dominantly present in the weathered soils. These soils are present in the tropics. There will be no major temperature limitations in these soils and they are acidic with minimal fertility rates. The soils in this region can be dominated by entisols, ultisols and oxisols. Oxisols are in high proportions in the soils. These soils will have minimal soil fertility and a high rate of Al saturation. The soils that are deficient in calcium ions can exhibit minimal cation exchange capacity. Because of this, there will be increased leaching in the soils. Due to leaching, most of the calcium ions will be lost. Due to Al^{3+} ions in the soils, the plants might not grow well. The calcium-based chemicals can be supplied to the soils that are acidic in nature. The increased calcium ions and the addition of gypsum in the soils can encourage calcium absorption in the plants. Gypsum can be added with lime and supplied to the soils. Due to this, there will be leaching of sulfates and calcium ions resulting in enhanced root growth.

Calcium ions can enter the lamella of the cell wall. It is important to know that the new plant parts may not receive calcium ions from the older ones.

Figure 2.6: Calcium deficiency.

Source: https://www.haifa-group.com/online-expert/deficiency-pro/ca-calcium

2.6 MAGNESIUM

Magnesium (Figure 2.7) can be classified as an important macronutrient in the growth and development of plants. The magnesium deficiency can dominate the soils with coarse texture. This is common in humid regions. The soils with low CEC can be deficient in magnesium ions. The minimal concentration of magnesium in the soils can be linked with the leaching process in the soil profile. In the case of sandy soils, the Mg^{2+} concentration is about 0.5 gm/kg and in the clay soils, the concentration is about 5gm/kg (Mengel et al., 2001). The availability of magnesium ions can be regulated by several factors. Important sources of magnesium are weathering, green manures, crop residues, and liming. The loss of magnesium from the soils can be due to leaching, soil erosion, and plant absorption. Some of the magnesium ions are fixed by the microbes as soil colloids. The cation affinity can be linked with the colloid properties. The magnesium ions that are present in the soils can be classified as water-soluble, exchangeable, and non-exchangeable or soil solution. The magnesium ions can be transported through diffusion and mass flow. Magnesium can be utilized by the plants in the enzyme activation process.

Figure 2.7: Magnesium deficiency.

Source: https://www.bioleaf.co.nz/articles/nutrients/magnesium-deficiency-in-plants

2.7 SULFUR

Sulfur (Figure 2.8) can be considered as an important macronutrient. Most of the crops respond to sulfur concentration. The soils that are deficient in sulfur were reported in several countries namely Thailand, India, Bangladesh, Brazil, and Indonesia. The limited presence of organic matter in the soil can deprive the sulfur deposits. Due to unfavorable conditions existing in the environment, there will be low rates of mineralization and this will lead to decreased sulfur concentrations.

The high-yielding varieties and intensive cultivation can extract most of the sulfur deposits from the soils. The use of fertilizers without sulfur can also leave the soils deficient in S. Soil erosion can be considered as an important factor in the deprivation of the S in the soils. The K and N fertilizers need high S concentrations. The fungicides and pesticides that have low S concentrations can lead to a reduced supply of S to the soils. In industries, if sulfur dioxide emission is reduced, there will be limited supply to the soils. The S molecules in the soils can be due to leaching and volatilization. Some of the humid and sub-humid climates can naturally reduce the S content in the soils.

Figure 2.8: Sulfur deficiency.

Source: https://brownfieldagnews.com/news/sulfur-deficiency-in-corn-detected-in-dry-areas/

Sulfur can be supplied via sulfur gases in the atmosphere, soil minerals, and organic matter. Some portion of the irrigation water can supply S molecules to the soils. According to the International Rice Research Institute, the flooded rice can sustain at least 54% of S and this was due to irrigation water (Tisdale et al., 2015). It is important to use sulfur-based fertilizers according to the plant needs. It is important to know that S-deficient soils will get enhanced and this can be attributed to the implementation of pollution control measures, expensive chemical fertilizers, use of S- free fertilizers, and cultivation of high-yielding varieties that consume more S.

The biochemical reactions linked with sulfur should be studied in detail so that we can proceed with the corrective measures needed in case of nutrient deficit conditions. The sulfur molecules can be added to the soils through crop residues, farmyard manures, and fertilizers. The sulfur dioxide and hydrogen sulfide gases are released into the atmosphere and will be brought down via rainfall. The sulfur dioxide gas will get reacted with water molecules and there will be the formation of acid rain. The sulfur molecules will be present in the inorganic and organic forms. We can observe an organic form of sulfur in the soils. The organic form of sulfur that is present in the soils can be classified into residual sulfur, C-bonded sulfur, and ester sulfate. In the case of amino acids, we can observe C-bonded sulfur. The amino acids like methionine and cysteine have C-S bonds. About 31% of the organic sulfur present in the soils comprise of methionine and cysteine. Most of the organic S forms can be mineralized by the soil microbes and then they can be absorbed by the plants. The sulfur compounds can be oxidized via biological and chemical processes. In many developing countries, the sulfur molecules will be supplied to the soils via natural phenomena like the rainfall that can bring down the sulfur dioxide molecules. The sulfur molecules that are present in the atmosphere can be an important factor in determining the variability in the temperate and humid regions. The sulfur concentration in the temperate soils will be high. The sulfur molecules in their inorganic form accountfor 10% of the sulfur present in the soils (Barber, 1995). The inorganic S can be in the form of H_2S, $FeS_{2,}$ and FeS. Thiobacillus bacteria can oxidize elemental sulfur.

2.8 ZINC

The importance of zinc in the plants was observed by Sommer and Lipman in 1926 (Marschner, 1993). Zinc can be classified as a micronutrient as it is needed by the crops in minute amounts or trace amounts. It can affect crop yields all over the world. The production of cereals in half of the soils all over the world is based on zinc concentrations. The zinc deficiency was prevalent in the corn crops.

Zinc deficiency (Figure 2.9) has been found in wheat, soybean, corn, and rice. Due to zinc deficiency, there will be reduced grain yield and the quality of grain will be poor. The pH of the soils will be increased due to liming as alkaline soil particles can absorb the zinc molecules. Increased levels of copper, manganese, iron, and phosphorus can enhance zinc deficiency in crop plants. This deficiency can be due to antagonistic interactions. Soil erosion can be considered as the main factor for the loss of zinc in the topsoils. Low organic matter in the soils can be considered as an important factor that dictates zinc deficiency.

Figure 2.9: Zinc deficiency.

Source: http://www.eplants.com.au/zinc-deficiency

There was a positive correlation between organic matter in the soils and zinc concentrations (Alloway, 2004). In the case of calcareous soils, zinc deficiency is high. Some of the minerals that have zinc content are hemimorphite, smithsonite, and sphalerite. Zinc is considered as an immobile element and this nutrient can be supplied to all plant parts via diffusion. The chelating agents can be produced by the microbes and root exudates will play a vital role in the transport of zinc (Stevenson, 2015). The zinc content in the soils can be entrapped in the minerals, may be occluded in the hydroxides and oxides of manganese, iron, and aluminum can be adhered with the organic matter, may react with the soil components, and can exist in the soil solution. The zinc molecules that are present in the soil minerals may not be absorbed by the plants. The absorption of zinc molecules can be affected by carbonate content, aluminum oxide, and iron oxide, organic matter, clay minerals, and pH of the soils.

2.9 COPPER

The importance of copper (Figure 2.10) in plant metabolism was observed by Lipman and MacKinney. Most of the forage legumes, grasses, vegetables, and cereals have a copper deficiency. The sandy soils that support minimum organic matter can be deficient in copper. If crop recycling is not practiced, the soils will be copper deficient. Anthropogenic and natural factors can limit the availability of copper molecules to the plants. Intensive cultivation and use of high-yielding varieties can consume most of the copper content from the soils. The loss of the topsoil can result in the removal of copper residues. Sometimes, the copper deficiency can be due to the utilization of herbicides that can be applied to the soils and leaves. High pH in the soils can encourage copper dissolution.

Figure 2.10: Copper deficiency.

Source: https://keys.lucidcentral.org/keys/sweetpotato/key/Sweetpotato%20Diagnotes/Media/Html/TheProblems/MineralDeficiencies/CopperDeficiency/Cu%20deficiency5.jpg

Some of the factors that can affect copper concentration in the soil are crop infestation, soil texture, soil CEC, root morphology, the redox potential of the soils, imbalance of micro and macronutrients, parent material, microbes, organic matter, moisture, temperature, crop species and pH of the soils. Copper can be considered as an important entity in the oxidase enzymes. Some of the important oxidase enzymes are lactase, ascorbic acid oxidase, and cytochrome oxidase. This chemical element is important in peroxidases and plastocyanins. It is involved in the detoxification of oxygen radicals that are produced in the photorespiration process. The copper deficiency in the legumes can result in lowered nitrogen fixation and nodulation.

2.10 IRON

Iron (Figure 2.11) is important in the growth and development of plants. It is a main constituent of the plant metabolic systems. Most of the enzymes that are involved in metabolic activities have iron molecules. The absorption of nutrients will be increased if the iron content increases. The toxicity and deficiency of iron molecules can be important in crop production. Chlorosis occurs in plants if the iron deficiency is high. Several factors control iron deficiency in plants (Wiersma, 2005). The root infection, imbalance in CEC, minimal aeration, temperature abnormalities, and use of insufficient iron fertilizers can cause iron deficiency. In the oxisols, iron deficiency can be observed in the beans and rice plants. The main reason behind it was that the pH value was> 6.5. Important sources of iron are parent materials, organic manures, crop residues, inorganic fertilizers, and other natural sources. Most of the iron molecules in the soils will be lost due to microbes, soil erosion and the excess uptake of iron molecules by the high-yielding varieties. The solubility and availability of the iron molecules can be important in determining the iron interactions with plants (Follett et al., 1989). Most of the iron molecules are classified as insoluble in the soil. Some of the iron-based minerals are magnetite, maghemite, lepidocrocite, hematite, and goethite. Ferric hydroxide can be considered as an important molecule that can supply iron molecules to plants. The minerals will weather and decompose to get precipitated as ferric hydroxides and ferric oxides.

Figure 2.11: Iron deficiency.

Source:https://keys.lucidcentral.org/keys/sweetpotato/key/Sweetpotato%20 Diagnotes/Media/Html/TheProblems/MineralDeficiencies/IronDeficiency/ Fe%20deficiency1.jpg

2.11 MANGANESE

Manganese (Figure 2.12) can be classified as an important plant nutrient in plant metabolism. The toxicity and deficiency of manganese molecules were reported in many parts of the world recently. ¼th of the soils suffer from manganese deficiency. Most of the manganese deficiency can be found in the coastal plains that are poorly drained and having coarse texture. This manganese deficiency can be based on several factors.

They can be climatic factors, soil, and plant-based factors. Some of the other factors that can contribute to manganese deficiency are crop species, soil type, root system, organic matter, pH of the soils, aeration, moisture, and temperature of the soils. The main sources of manganese are microbes, chemical fertilizers, and organic manures. Manganese can produce hydrated oxides.

Figure 2.12: Manganese deficiency.

Source: https://www.agric.wa.gov.au/sites/gateway/files/Manganese%20deficiency%20in%20tomato%202%20%28NSW%20DPI%29.jpg

The redox potential of the soils can affect the availability and solubility of the manganese compounds in the soil systems. Fungi and bacteria can affect the cycling of manganese. Mn has many functions in plant metabolisms. Mn can help the plants in producing lignin, glycosides, and phenolics. Mn can be associated with the nitrogen fixation process in the legumes. The fixation of carbon dioxide in the C_4 plants can be associated with the manganese concentration. The oil percentage in the seeds may get decreased due to limited Mn in the plants. Mn can protect the plants from several diseases. The plants that have Mn deficiency can become sensitive to the freezing process. Mn deficiency can reduce the carbohydrate concentration in the seeds and fruits. The assimilation of nitrates and iron metabolism can be linked with Mn deficiency. Mn is one of the components of several metalloproteins. Most of the redox reactions and electron transport systems can be linked with the Mn concentrations. The production of oxygen in the plants can be affected by Mn. It plays an important role in the germination of the seeds. Most of the respiratory enzymes in the plants need Mn. Chlorophyll can be produced with the aid of manganese molecules. The younger leaves face manganese deficiency. The stems can become weaker and the leaves will become shorter.

2.12 NICKEL

Nickel or Ni can be considered as an important micronutrient that can be found mainly in the higher plants. Ni was found in the plant ash. Ni when sprayed on leaves enhanced the crop yields of beans, potatoes, and wheat. The plants that are Ni deficient can absorb more urea (Shaddox et al., 2017). Ni was classified as heavy metal and if this element is present in higher concentrations in the soils, we can call them as contaminated soils. Ni toxicity can be observed in most of soils. The nickel molecules can be added to the soils when the parent materials are weathered. This can be due to municipality compost, sewage sludge, and acidic soils. The depletion of Ni can be due to plant absorption, microbes, soil colloids, and soil erosion. Ni deficiency can be observed in plants but not frequently. The deposition of Ni on the minerals can occur due to precipitation and adsorption. Nickel can be classified as a mobile element and so the symptoms appear in the old leaves than younger leaves. The nickel deficiency can be studied by symptoms like reduced seeds, early senescence, and chlorosis.

2.13 CHLORINE

Chlorine or chloride ion is one of the micronutrients but the absorption amounts of this nutrient are high and are almost the same as a macronutrient. This element can be classified as non-metal. Chlorine deficiency can be observed in the sugarcane crops. Chlorides can enhance the grain yields of wheat. Chlorides can be added to the soils through air pollution, fertilizers, rain and in some case soil reserves. The chloride ions can be distributed due to recycling and there will be both depletion and addition of Cl^- ions. The chloride ions are important in the photosynthesis mechanisms. Due to chlorine deficiency, the transpiration mechanisms will be altered. The increase in crop yields can be based on many factors like plant-water relations and nutrition (Fixen et al., 2018). Chloride ions can be important in osmoregulation. Due to the reduced supply of chloride ions in the plant systems, we can observe the leaf spots. There will be osmotic imbalance due to chlorine deficiency. In wheat, the leaf spot syndrome is prevalent due to chlorine deficiency. The chlorine molecules can affect plant nutrition as the chlorine deficiency can lower the nitrification process. Chlorine deficiency can lead to chlorosis. The dry weight of the plants will be reduced. The uptake and concentration of the chloride ions in the crop plants can be based on the climatic conditions, crop yield, and soil solution. The chloride ions can be classified as mobile chemicals and so the availability of the ions to the plants can be easier. The negative charge on the chloride ions will be sustained in the plants. The fluctuation in the Cl concentration in the soil solution can be important as it determines the availability of chlorides for plant absorption.

2.14 MOLYBDENUM

Molybdenum (Figure 2.13) or Mo can be vital in plant growth and development. It is needed by plants in trace amounts. This element is important in the enzymatic reactions that occur in plant metabolism.

Figure 2.13: Molybdenum deficiency.

Source:https://www.canr.msu.edu/news/molybdenum-deficiency-in-poinsettia

The Mo concentration in the soils is low and it occurs in the soil solution. Mo can be adsorbed on the iron hydroxides and oxides. The availability of Mo will be more if the pH value increases. There will be increased Mo absorption by the plants if the pH value is greater than 6.5 (Jones, 1952). The Molybdenum ions that are present in the soils can exist as oxyanion or MoO_4^{2-}. The Mo ions can exist in many oxidation states but they are less in the agricultural soils. The Mo ions can be accumulated in the organic matter. The reduction of nitrate N into nitrite N can be catalyzed by Mo ions. The Mo ions can be important in the formation of nodules. It helps in the symbiotic N fixation. The Mo ions are present in the nitrogenase and nitrate reductase. The Mo ions can reduce wheat sprouting and this can be due to increased abscisic acid. Mo can decrease the incidence of diseases. The Mo deficiency can occur in the older leaves. In some cases, we can observe the Mo deficiency in the younger leaves. Some of the symptoms due to Mo deficiency can be the same as those observed in the calcium-deficient conditions. There will be marginal chlorosis of leaves. There will be stunted plant growth. We can observe narrow leaf margins and mottled leaves. The toxicity of Mo cannot be mimicked in the field conditions. Molybdate ions can be present in acidic soils with a pH less than 4. The plants can absorb the molybdate ions and the diffusion can take place through the roots. Mo can get transported via xylem and as anions. Due to increased Mo levels, the iron translocation will be affected. There will be apical chlorosis and stunting in the plants with high Mo levels. P in its soluble form can increase Mo uptake. The Mo and Mn levels in the leaf tissues will exhibit an inverse relationship. The application of N can get lowered due to absorption of Mo. The uptake of Mo can decrease with soluble S. There will be antagonism between copper and molybdenum. The liming of the acid soils can enhance Mo absorption. It is important to do soil tests frequently so that the rate at which Mo ions are being absorbed by the crop plants can be evaluated. It was observed that the Mo deficiency was rarely found in the calcareous and neutral soils. We can reduce Mo deficiency by the addition of fertilizers. Sodium molybdates and ammonium molybdates can be classified as soluble compounds that can be used as Mo fertilizers.

2.15 BORON

Boron can be classified as an important micronutrient for plants. It was classified as a non-metal element. It mostly exists as a neutral molecule. The absorption of B by the plants is low, but still, B deficiency is observed all over the world. Natural deficiency and induced deficiency are the two types of deficiencies observed in the case of Boron. The natural deficiency can be due to decreased production of B from the parent minerals. Induced deficiency can be due to several environmental factors. It is important to know the B recycling in the soil and plant systems so that there will be effective nutrient management. The major sources of Bare parent materials, organic matter, and chemical fertilizers. Most of the B can be removed via leaching, microbes, adsorption, and crop uptake. The salinity of the soils, organic matter, clay content, cation exchange capacity and soil pH can be considered as important factors that can affect the B concentrations in the soil.

Figure 2.14: BORON DEFICIENCY.

Source:https://www.notcutts.co.uk/garden-advice/problems-pests/boron-deficiency/

Kaolinite can be considered as a least reactive substance with B and illite can react with B (Keren et al., 1958). Boron can be linked with the production of pectin in the cell wall. It also participates in cell wall growth. B can enhance seed vigor and germination. The P and N metabolism can be affected by B concentration. The elongation and development of the cells can be affected by B. The number of pods in the legumes can be enhanced by the addition of boron. The abortion of pods in the legumes can be prevented by B application. B can play a vital role in the transport of nutrients. The translocation of the sugar molecules can be regulated by B. If Boron is reduced in plants, the nitrates will get accumulated in the stems, leaves, and roots. The synthesis of nucleic acids can be catalyzed by Boron molecules. There will be reduced production of cytokinins due to B deficiency. The formation of cell walls and seeds can be controlled by B molecules. B plays an important role in development of the pollen tubes and pollen grains. Boron was classified as an immobile element. This is the reason why the symptoms due to B deficiency appear in the younger leaves. There will be a reduced distance between the internodes. The flowering rate will be reduced in the plants with B deficiency (Dordas, 2006). There will be a crackling of the petioles due to boron deficiency.

Chapter 3

Nutrient Efficiency and Management

CONTENTS

3.1 Introduction 52
3.2 Physiology 54
3.3 Nutrient Use Efficiency 57
3.4 Nue For Macronutrients 63

3.1 INTRODUCTION

It was observed that the plants require C, H and O in larger amounts and they absorb these elements from the soil and atmosphere. Some of the mineral elements like chloride, sulfur, potassium and phosphorus will be extracted from the soil solution via roots. Most of these mineral elements were found in the plant ash. The supply of nutrients from the organic manures is an essential task than supplying the organic matter itself. Most of the organic manures contain K ions and they were classified as water-soluble elements. It is important to know that the K ions get retained because of the soil organic matter. The organic matter can offer cation exchange sites. Animal manures are classified as important as far as the crop plants are considered. The slurry and the Farm Yard Manure can be considered as animal manures. These manures contain most of the P ions. The soil particles and granules can hold the P ions and they are equivalent to the ions obtained from the mineral fertilizers. It is important to cleave the organic phosphates into inorganic phosphates so that the plants can absorb the inorganic P ions readily. It was proved that P can be easily replenishable and will be available to plants easily. Nitrogen is also an important element that controls the growth of the crops. The nitrogen supplies to the crop plants have enhanced crop growth drastically. It is always hard to maintain supply and demand in equilibrium as far as the N fertilizers are concerned. The soil moisture and activity of the microbes can be important in determining crop growth. Fertigation can be classified as an important method in maintaining plant nutrients. The application of fertilizers can be through the foliar route and it is essential in the cropping seasons as the plants would have grown well. It is important to check for mineral deficiency so that crop quality can be ensured. Several experiments were carried out to establish the efficiency of foliar application and fertigation. (Figure 3.1).

It has become a common practice to select the plant materials that will suit the local soil conditions. It is important to add large quantities of fertilizers. The fertilizers can enhance crop yields. It should be kept in mind that getting the enhanced yields in the P deficient soils cannot be associated or linked with the enhanced P efficiency. Most of the growth patterns will be affected by the genotypes of the plant species. The yield produced per unit of P used can be defined as the nutrient efficiency of P of that crop.

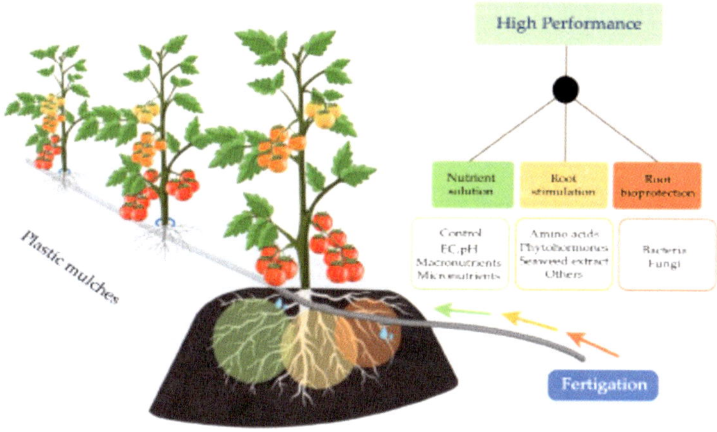

Figure 3.1: Fertigation.

Source:https://www.mdpi.com/2223-7747/7/4/88/htm

The nutrient uptake by the plants can be improved by several factors. It is based on the ability of the plants to take up the nutrients that are less available to them relatively. It is also based on the nutrient uptake per root cell or the roots. The contact of roots with the nutrients can be an important factor in determining the nutrient uptake. The uptake of P can be linked with VAM fungi (Figure 3.2) or Mycorrhizal fungi. The assimilation of N molecules in the plants can be due to Rhizobium bacteria. The enhanced size of the roots can enhance these nutrient uptake. The size of the roots can differ with the plant species. These differences can lead to varied nutrient uptakes. The hybrid varieties can absorb more nutrients than the wild genotypes. The size of the root and the growth rate of the plant can determine the nutrient uptake.

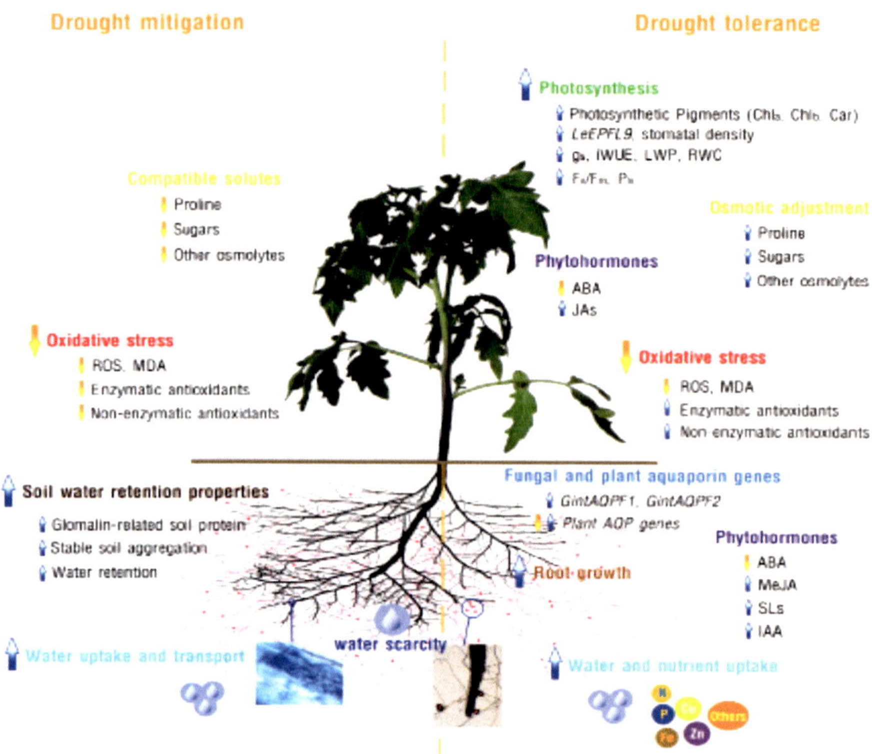

Figure 3.2: Benefits ofVAM fungi.

Source: https://www.intechopen.com/books/drought-detection-and-solutions/benefits-of-arbuscular-mycorrhizal-fungi-application-to-crop-production-under-water-scarcity

3.2 PHYSIOLOGY

The plants can be in diverse environmental conditions. This can be due to the varied utility of solar energy. Solar energy can be considered as an important factor that drives the chemical exchange between the biotic and abiotic factors of the environment. It was evident that 13 nutrients with C, H, O are important in plant metabolic activities. The nutrients can be of two groups. They are macronutrients and micronutrients. These nutrients will play important role in the determining physicochemical properties of the plants. The nutrient use efficiency can be considered as a complex system and it is controlled by multiple traits. Most of the physiological processes that regulate nutrient use efficiency are interconnected. These processes are dependent on several factors. Breeding can be used in the development of cultivars and species that are high-yielding varieties. There will be no equilibrium between the environment and plant systems. There will be a synchronization of physiological reactions, morphology, development, and life cycle in the plant systems. The absorption and assimilation of the nutrients can be important. The nutrient use efficiency can be evaluated in diverse populations, ecosystems, communities, and species. Agroecology studies that determine the association between the biotic and abiotic interactions can be important in understanding the nutrient use efficiency. The nutrient use efficiency or NUE can be inversely proportional to the nutrient concentration in the tissues. Some of the nutrients can be absorbed from the atmosphere.

Most of the plants are dependent on the mineral nutrients that are present in the soils. The mineral nutrients can be formed due to natural weathering of the parent material or in some cases, biological intervention that can convert the atmospheric elements into organic substitutes. The acidification of the soils can enhance the uptake of nitrates in the rhizosphere. The storage of nutrients can be considered as an important process that includes absorption, the formation of reserves and recycling. The increase of the compounds can be considered as an accumulation step and it is not considered as an important factor in the direct developmental process. The formation of reserve material can be a controlled process that helps metabolism. The CO_2 present in the atmosphere will be converted into carbohydrates through photosynthesis. The photosynthesis process can be considered as an important source of biomass production. N can be associated with the photosynthesis process directly. All the important nutrients along with the photosynthesis process can enhance biomass production. Photorespiration can be considered as an important process in plant metabolism (Kozaki et al., 2014).

The N molecules can adhere to the carbon molecules and so it is transported to the leaves. The decomposability of litter and the re-sorption of the nutrients must be studied in detail. The translocation of N can determine the quality of the grains and with that, the nutrient use efficiency will be increased (Rose et al., 2010). The loss of nutrients can be an important factor to be considered whilst studying nutrient use efficiency. The nutrients will be lost via many paths and often we may not be equipped to know all the routes. The loss of nutrients from the leaf surface can be due to volatility and leaching (Eichert et al., 2012). The senescence or litterfall can be considered as an important method of losing nutrients. The physiological process in determining the nutrient use efficiency can be a complex process and it is based on several factors. In order to understand nutrient use efficiency or NUE, we need to understand that environment plays a major role. The crop yields will be based on the environmental conditions and the effect of the environment on the crops is predominant in the growing period. Some of the environmental factors are soil composition, rain, light, and temperature. It is always important to enhance crop production than to focus on the NUE. Drought can be classified as an important factor that is responsible for nutrient loss in many plants. The mass flow of water molecules can be associated with the transport of nutrients in the plant systems. The nutrient loss that occurred in the maize plants was compensated with the supply of N. Whilst drought stress, the transportation of the essential nutrients via xylem will be impaired. Slight stress due to drought conditions can enhance the phosphorus deficiency in the crop plants. The uptake of nutrients and growth rate can be declined in the light limiting conditions in nutrient-rich environments (Caplin, 2018). Due to extreme temperature conditions, the uptake of the nutrients will be reduced or slows down. Generally, the nutrient cycles and the mineralization process will be based on the properties of the soil. In agroecosystem and ecosystem, the litter composition will play an important role in nutrient cycling. The deposition of litter and its decomposition is based on temperature conditions. The weathering rate and age of the parent material can determine the nutrient capacity of the soils. The rhizobacteria and arbuscular mycorrhiza can enhance the surface area of the roots so that the absorption rates will be enhanced. The gas traces will get altered in the plant systems. There are some gases that can affect plant metabolism. Most of the S and N compounds can be classified as pollutants and these elements are dumped in the atmosphere due to the anthropogenic intervention. There are certain extreme events like volcanic activities that can bring about these elements. There are some elements if present in higher concentrations can be toxic to the plants. Nitrogen can be absorbed by the plants in the form of ammonium ions that are positively charged. It can be observed that the calcium and magnesium cations will be disturbed and this can help in managing the charge balance. The crop plants that were sprayed with the nitrate fertilizers can take up less amount of phosphates. In the root medium, it was found that either nitrate or sulfate was accumulated (Koralewska et al., 2015). It was evident that if the vacuoles are previously filled with the nutrients of charge, the other charged nutrients may not occupy the vacuoles due to space requirements. There were certain interactions with other nutrients if there are several other nutrients with diverse charges. We can find specific interactions between iron and sulfate nutrients (Musetti et al., 2012). There will be the production of Fe-S clusters (Forieri et al., 2016). The interactive effects between ammonium, nitrates and sulfates were observed. In the case of high S content in the atmosphere, the rhizosphere area will have separate nitrate and sulfate absorption instead of coupled absorption. The transport of sulfate and phosphate in the membrane of the chloroplasts can be classified as competitive. The important problem whilst studying nutrient use efficiency is the non-stomatal and stomatal absorption. The climate change conditions have forced the crops to adapt with extreme conditions and water use efficiency can be improved (Reynolds et al., 2011). In the case of drought conditions, it is important to select plants with improved root traits. The enhanced supply of nitrogen molecules can increase water use efficiency. It was observed that the light use efficiency and nitrogen absorption are linked. There will be certain genetic adaptations that can be observed in the plant species with traits of ecological and habitat advantage. In the case of novel breeding programs, it is important to improve nutrient use efficiency and yield potential. The transgenic crop plants can be used to enhance the NUE.

3.3 NUTRIENT USE EFFICIENCY

The plants can be adapted to most of the climatic conditions even if the optimal conditions are disturbed and extreme weather events occur. It was observed that plants can inflict certain integrated responses to the rough weather conditions. Plants exhibit quantitative variation that can be quantitative in nature. The phenotypic variation can occur due to many genes and they can be used in most of the breeding programs. Most of these programs are aimed at enhancing the quality and yield of the crop plants (Bergelson et al., 2011). Sulfur, phosphorus, potassium, and nitrogen can be present in large amounts. It is important to maintain an equilibrium in the plant systems so that all these four elements are available in good proportions. These elements can increase the crop yield, taste, and nutritional quality with enhanced resistance against several pests. There are some plant species that can grow in soils with fewer nutrients and water. It is important to know the mechanisms that can regulate the absorption of nutrients. Natural variation can exist between the plant species and it can be explained in terms of maturity age, flowering age, seed dormancy, the composition of the grains and leaf shape. Natural variations are important factors in determining the quality of the crop products (Robinson et al., 2008). Plant plasticity plays an important role in determining the adaptability of the crop species. This is based on many environmental factors. The age of the plant and its plasticity can be important in the breeding programs. The physiology and morphology of the plant species based on several quantitative traits. It can be said that most of the characters in the plant species against environmental constraints can be due to their genetic setup. Some alleles can be detected with the mutant screens. It is always important to study natural variation as a function of an evolutionary perspective. The molecular and phenotypic variation are important in understanding the adaptive nature of the plant species. The nutrient uptake efficiency can be based on the composition of the root system. The plasticity and the root composition can be used to describe the nutrients that are scarce in nature (Kellermeier et al., 2013). The genomic studies explains the assimilation of the nutrients in the plant species. The nutrient content in the seeds can be used to determine the remobilization efficiency due to natural variation. There are some species in the Acacia genus that can produce vegetable fat and this can be attributed to the natural variations. There are many studies that helped us to describe the improvement of the crop yields and composition of the grains (He et al., 2011). The genetic approach to the quantitative traits can be called Quantitative trait loci (QTLs) mapping. The QTLs determine the response of the crops towards the environment. This technique of mapping is based on the genotyping of the offspring that is derived from the crossing of genotypes of different origins. The phenotypes produced will be compared with the molecular markers of the offspring. There will be the detection of certain genomic regions that are statistically significant. This significance will be based on trait variation and polymorphism. These regions are called QTLs. The analysis of the QTLs is used in determining the natural variation in the species under study. If the genetic variation is found, then it is important to know the location of the loci in the genome. There are certain steps that are to be followed in the detection of the number of QTLs (Figure 3.3) and their position in the genomes. It is important to have a mapping population for the experiment. The genotyping of the markers should be done for the complete genome. The phenotyping of a particular trait must be done. Then, the association analysis of the genotypic classes and the phenotypic value of a particular trait of the markers that are polymorphic must be done. The genetic position and number of loci that may regulate the variation of a trait in a sampling population must be studied.

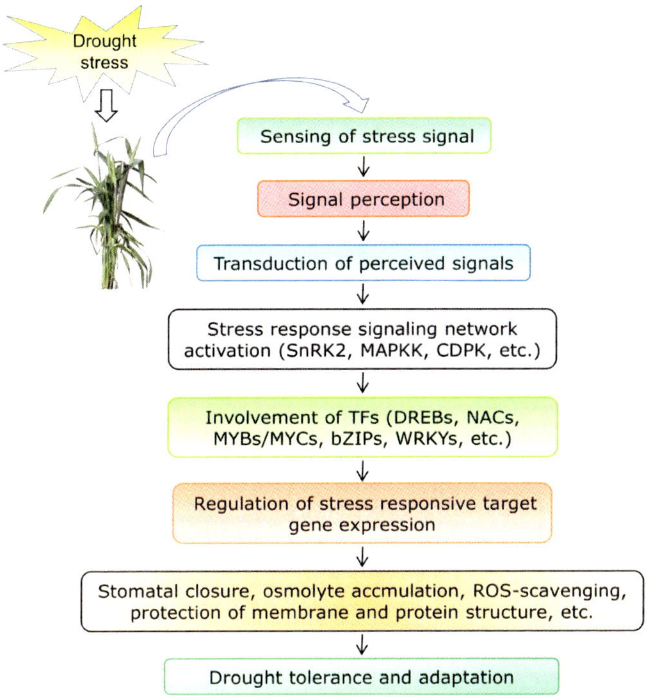

Figure 3.3: QTL.

Source:https://www.mdpi.com/2073-4395/7/1/5

The action mode of QTL for dominant effect and the interactions between loci for epistasis can be calculated for an experimental population. Generally, the number of loci in each analysis will be 1 or even greater than 10. It is mostly based on the genetic variation complexity. It is also based on the genetic interactions, an additive effect of QTL and segregation of loci. In selecting the experimental population, homozygous plants will be used. With this, we can easily perform multiple analyses and replications in the same population. These populations can be called RILs or recombinant inbred lines. (Figure 3.4).

In some cases, we can call them introgression lines or ILs. Most of them are homozygous in nature and so the environmental effects will be assumed to be negligible. These homozygous populations enhances the resolution of the QTL detection. We use this technique in different environments. The homozygous population are produced by selfing process. The Near Isogenic lines (NILs) and recombinant inbred lines (RILs) can be classified as common population types that can be used in the QTL analysis. The accuracy in this process can be based on the size of the population and this accuracy can be called mapping resolution.

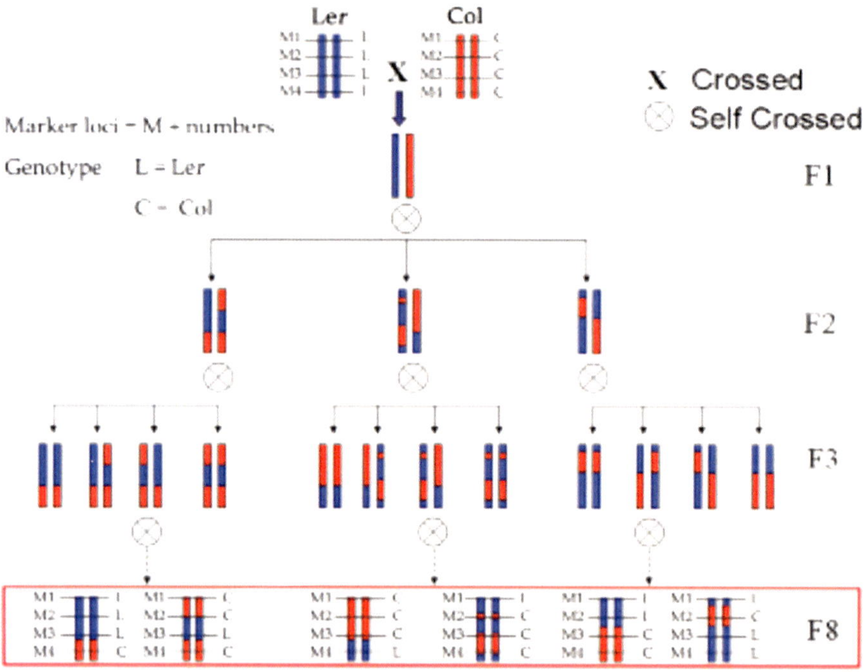

Figure 3.4: Recombinant inbred lines.

Source:https://www.researchgate.net/publication/36449375_Analysis_of_two_single_trait_loci_affecting_flavonol_glycoside_accumulations_in_Arabidopsis_thaliana_natural_variations/figures?lo=1&utm_source=google&utm_medium=organic

In the case of segregating populations, we can get reduced resolution. There are RIL populations, that can be used in the detection of traits like virus resistance, drought tolerance, germination of seeds and architecture of the roots. One of the important approaches in the QTL analysis is the use of multiple parents (Huang et al., 2011). The multiple advanced generation intercross can be called MAGIC population. This MAGIC method can produce several recombination events. QTL analysis was done for many enzyme activities in several crops. NILs can be used to map QTLs with finer resolution. The candidate genes can be co-localized and this will help us to perform QTL analysis and their molecular identification will become easier.

The microarrays and genome-wide techniques can be used to analyze the candidate genes. The QTLs can be validated only if allelic variation can affect the trait under study. The polymorphic regions-markers can be important in mapping QTLs in finer resolution. If the region mapped contains a smaller number of genes, then we need to assume that the QTL mapping is done at fine resolution and it is always important to use these genes for any clues. QTL cloning can be considered an important method in the validation of a novel gene. The analysis of QTL can produce large QTLs that are responsible for the production of fructose in the leaves. Once the data is procured from the multiple populations, it will be easy to know the QTLs for a specific trait. The QTL mapping can result in the detection of the variants that regulates the development and physiological traits. The species that can be used in this analysis can be self-fertilizing so, the strains gathered from the wild species are homozygous in nature. Genome-Wide association studies were done for several species as part of the genome project. About 1001 accessions were used in this project. The genetic mapping was done for many purposes in specific populations and the mapping was used to study the variations in the pedigrees. In the case of genetic association mapping, there will be a collection of individual samples that can comprise wild populations and their pedigree samples that were derived from the breeding programs. There are many alleles that can be evaluated at the same time in the case of a diverse population. Generally, there are 2 alleles that can segregate independently in the case of biparental offspring. If the knowledge of biochemical pathways is not sufficient, it is important to go for a genome scan (Rafalski, 2010).

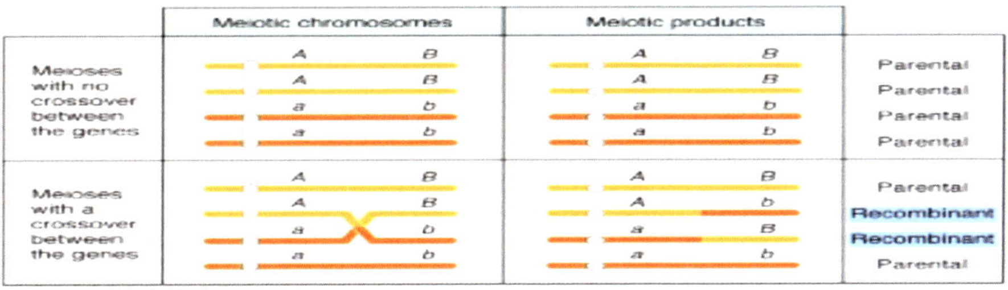

Figure 3.5: Linkage disequilibrium.

Source: https://www.slideshare.net/FAOoftheUN/mapping-and-applications-of-linkage-disequilibrium-and-association-mapping-in-crop-plants-55689150

In the case of genome scan, the testing will be done for many genome segments and they will be linked with the genotyping process. The genotyping will be performed on the loci of the genetic markers spread over the chromosomes. The genetic loci can be normal for a trait and in some cases, they will be LD or linkage disequilibrium. (Figure 3.5). The genetic loci are generally linked with each other. The germplasm of the maize crops can be split into heterotic groups. It is always important to understand the LD of a population in the research studies with the help of genetic markers. We can observe that the LD value gets diminished as the distance between the loci increases. This occurs at a slow rate in the case of soybean inbreds. In the case of maize outbreeds, this rate is faster. It was observed that these effects can impact the breeding methods (Flint et al., 2013). We observe that LD can be discrete along the genome. In the case of the *Arabidopsis* genus, LD can be 10kb and this can be considered as sufficient as far as the genetic mapping is concerned. There are high throughput tools that can help us in precision phenotyping. This phenotyping can be called phenomics. This technique will be based on automation (Finkel, 2019). This phenomics technique is being used in the analyses of plant responses to various stresses (Tisne et al., 2013). This technique was used to identify the QTLs that were linked with the drought response was mapped. It must be considered that hypotheses be validated in the case of association mapping. The backcrossing process was repeated in several experiments. The biparental populations were used to understand the locus and relevant alleles. The breeders should identify the alleles at loci and it will be depending on the mapping attempt and the markers being used in the selection of individuals (Ribaut et al., 2010). These approaches can help us in fixing the alleles of interest in the desired populations.

There are some limitations to this technique. The association mapping on the relative loci and the magnitude of the locus in the population. It is also based on the frequency distribution of the alleles. It is important to know that this technique is not good at determining the alleles of rare occurrence. It is important to know that the segregation of the biparental populations is important in mapping the alleles. The enhancement of the phenotyping quality can be considered an important task as far as the traits are concerned. There will be natural variation within the species and it is evident in the plant species. The mapping can be classified into QTL mapping and association mapping. In the case of QTL mapping, there will be a choice to select the parental lines. The offspring must segregate the desired trait. In the case of association mapping, the genetic composition of the desired trait may be known beforehand. If the genetic composition is known, then we can search for the candidate genes. If the genetic composition is not known, then we need to go for genome-wide association mapping. In the case of QTL mapping, we need to produce a suitable population and they may be MAGIC, RILs and F_2. In this QTL mapping, there will be a link between the polymorphic markers and phenotypic values.

Then, we can detect QTLs. There will be a validation of QTLs. Then, we need to go for the candidate gene approach. Finally, the desired genes that control the traits will be identified. After validation of QTL, there will be fine mapping. In the case of association mapping, if the genetic composition is not known, we need to test for a link between the phenotype and the polymorphic markers that are located all over the genome. If the genetic composition is known, we need to check for polymorphic markers in the genes and they can be derived from various genotypes. Finally, there will be a test for a link between the phenotypes and the makers. Later, we can identify the genes of desired traits.

3.4 NUE FOR MACRONUTRIENTS

It was reported that sulfur supply must be needed for a good rate of growth in the plant species. This condition applies to almost all organisms. Sulfur is present in most of the metabolites and it is important in several biochemical reactions. Most of the sulfur content will be recycled as part of the global ecosystem. The inorganic sulfur will be converted into organic sulfur. The conversion can be through microbes and photosynthetic organisms. SO_4^{2-} can be considered a natural form of sulfur. It will occur in the oxidized form. S molecules can exist in the +VI state. The plants absorb the sulfur molecules from the soils in their reduced form and this process can be classified as energy- dependent process. Sulfides can be utilized by the plant species to produce cysteine amino acid. This amino acid can be stable. S can be present in the cell walls, sulfolipids and methionine. It is also present in the plant cells and will participate in the redox reactions. The coenzymes, vitamins can comprise sulfur compounds. Alliins can be isolated from garlic. The degraded products of the glucosinolate can activate enzymes that can reduce the formation of the tumors. In European countries, the sulfur content in the soils was declining. The use of high-quality fertilizers with almost no S content deprived the soils. The declined emission of SO_2 has also affected the S content in the soils. The plant breeding programs and intensive agriculture has enhanced the demand for S in the soils. In order to produce 1 tonne of oilseed rape seeds, it is essential to use 16 kilos of S (McGrath et al., 2006). One tonne of wheat may need 3 kilos of S (Zhao et al., 1999). This disparity in the S absorption and utilization can be attributed to the xylem and phloem transport mechanisms. The oilseed rape is much sensitive to the S deficiency. The S concentration can determine the baking quality of wheat. The bread quality will be based on the proteins present in the wheat flour. Glutenins and gliadins that are classified as gluten proteins are present in large amounts. The extensibility and elasticity of the flour can be based on the S concentrations and proteins associated with it (Hussain et al., 2013).

Due to S deficiency, there will be undesirable variations in the wheat quality. The acrylamide formation in the wheat and potato-based products was considered an important problem. The concentration of the free asparagine can be a determinant of the acrylamide. In the case of S-deficient conditions, there will be more accumulation of free asparagine in grains. Due to this there will be an increase of asparagine in the amino acid pool. This forces us to use sulfur-based fertilizers on fields so that the formation of acrylamide will be checked effectively (Halford et al., 2001). The intake of the sulfates is from the soils. Once sulfates enter the plant systems, they get accumulated in the vacuoles and plastids. The sulfur transport proteins (Figure 3.6) play important roles in the transport of sulfates to all organs. The transporter gene family can be classified into five classes. The transporters that have high affinity can be important in the uptake of the sulfates from the soils. The cortex cells and the root epidermis have high affinity transporters. Most of these transporters will work if S is present in sufficient amounts. The activity of the transporters will be increased if the concentration of glutathione, cysteine, and sulfate is reduced. The status of S concentration in the soils may not affect the group-3 sulfate transporters. The groups one, two and three transporters can be inducible and semi-constitutive in nature. Group 4 transporters are important in the distribution of sulfate molecules. If the S nutrition is higher, then there will be repression of the genes that regulate the transporters. The entry of the sulfate molecules into the plant systems will be through the plasma membrane. The functioning and physiological effects of the transporters were well established. Most of the plants utilize the proton-sulfate system and this system was classified as the co-transport system that can catalyze the influx of sulfates.

Generally, there will be the establishment of a proton gradient. The transport mechanisms will be based on pH, sulfates, and protons. The transporters in the absorption of the sulfates were identified in the mutant yeasts. Due to this, yeast was considered an important tool in determining the function of the sulfate transporters. There are many sulfate transporters identified in plant species (Yoshimoto et al., 2003). The low-affinity transporters and high-affinity transporters were first identified with all the characteristics.

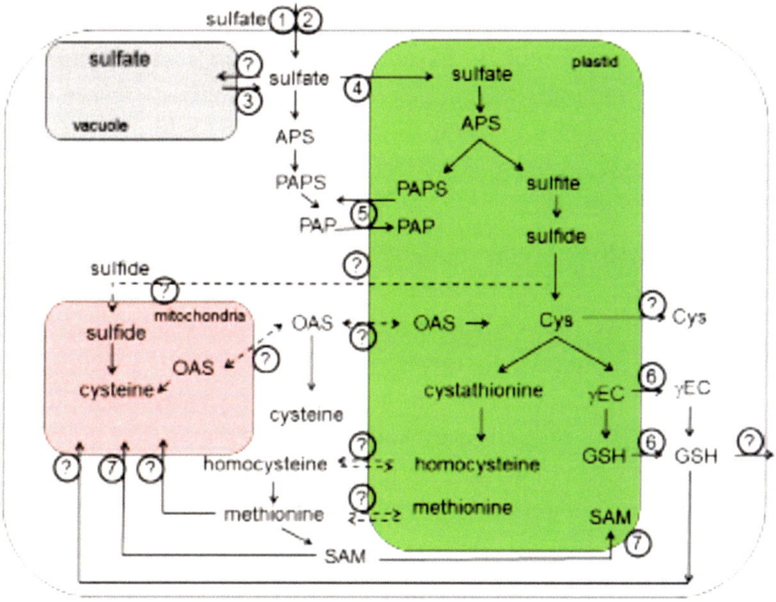

Figure 3.6: Sulfate transport proteins.

Source:https://www.frontiersin.org/articles/10.3389/fpls.2014.00442/full

The high-affinity transporters can enhance the uptake of the sulfates in the plant systems (Barberon et al., 2008). The absorption or assimilation of the sulfates in the roots can be through two transporters. They are SULTR 1;1 and SULTR1;2. These genes can be found in the cortical cells, epidermal cells, and hairs of the root system (Rae et al., 2002). The transport of the sulfate in the root system will follow the symplast mode of transport. SULTR1:2 can be classified as an important portion in the absorption of the sulfates.

The sulfates can get into the epidermal cells and they get transported into the xylem. It is always important as the vacuoles cannot store sulfates for long periods. Plasmodesmata can help in the transfer of sulfates from the epidermal cells into the central cylinder. In some cases, the sulfates can reach the apoplast. Through several studies, it was evident that the plants induce and activate certain additional genes in order to balance the sulfate distribution. The sulfates will get assimilated in the plastids and in some cases, they get accumulated in the vacuoles. So, it was considered that vacuoles can act as sulfate reservoirs. SULTR2;1 was identified in the chloroplasts. The growing plants need sulfates so that they can prepare the amino acids, thiols, and other compounds that can be used in the plant metabolism. The need for sulfur in the plants will be based on the developmental stage, organs, and tissues. The vacuoles that are present in the leaves can act as sulfate reservoirs. It is always important to know the distribution of the sulfates from the vacuoles to other plant parts. The regulation of the sulfates will start at the messenger RNA.

Sulfur can exist in diverse oxidation states. They can exist as bioorganic compounds, organic and inorganic compounds. The fungi, plants, bacteria, and algae can reduce the sulfates and can produce amino acids (Howarth et al., 2003). Whilst plant growth, the activity of the sulfates in the chloroplasts decline and their concentration enhances in the cytosol. In the case of *Euglena gracilis*, the reduction of sulfate will occur in the mitochondria. It is important to know that sulfates are stable. The stable sulfates must be activated prior to their reduction. This process occurs due to adenylation. As a result of adenylation, APS or Adenosine 5'phosphosulfate will be produced. This process will be mediated by ATP sulfurylase or ATPS. This enzyme will be about 54kDa (Murillo et al., 1995). ATPS will be accumulated in the plastids so that there will be activation of sulfates. There were reports that showed that the ATPS presence was in the cytosol (Rotte et al., 2000). The ATPS enzyme will be coded by several genes. There are two ATPS isoforms in the plants. In the case of the *Arabidopsis* species, there are four isoforms (Kopriva et al., 2011). There will be transit peptide in the chloroplasts. We can know the isoforms of ATPS by using start codons. APS can involve in the sulfate reduction mechanisms. APS reductase has a direct impact on sulfite reduction. The phosphorylation process can be mediated by APS kinase. There will be the formation of 3'-phosphoadenosine-5' phosphosulfate or PAPS. This PAPS can act as a donor for many reactions that can alter the production of secondary metabolites, saccharides, and proteins. This process can be called sulfation. This process is important in controlling the development and growth of plants. The reduction of sulfate can be considered as an important assimilation route in most plants. In this process, two electrons will be transferred to APS from APR. There will be the production of sulfites. Most of these electrons will be extracted from GSH. Sulfides will be formed from sulfites. This reaction will be mediated by sulfite reductase that is dependent on ferredoxin. About 6 electrons will be transferred to sulfite from ferredoxin. Then the sulfides formed will be added into OAS or O-acetylserine. From OAS, there will be the formation of cysteine. OAS thiol-lyase will mediate this reaction (Takahashi et al., 2010). In the sulfur assimilation pathway, APR can be considered as an important enzyme and this will be controlled by signaling molecules and environmental factors. There are many genes that can encode APR. The hormones have a direct effect on the APR2. PAPR needs glutaredoxin and thioredoxin as cofactors. The shift of phosphates from ATP can be mediated by APK enzymes. There will be a production of PAPS. The PAPS can be classified as a donor of sulfate. The enzyme sulfite oxidase mediates the oxidation reactions that can convert the sulfites to sulfates. This conversion can be classified as a 2e system and it occurs in the peroxisomes. The sulfite oxidase doesn't have a haem portion. In place of the haem domain, there will be the Mo domain. Molecular oxygen can be considered as an electron acceptor in the plant systems. In this reaction, the molecular oxygen will be converted into hydrogen peroxide. The sulfite can get oxidized by the action of hydrogen peroxide and there will be no enzyme involved. The sulfite oxidase or SO can protect the plants if the S concentration is high in the plant system. This enzyme also gets increased if the sulfur dioxide concentration is high in the atmosphere.

Cysteine can be considered as an important amino acid that helps in plant growth and development. The sulfides can get into the beta position of the serine molecule. This occurs in the last stage of assimilation of sulfur components. The serine molecules should be activated prior to the addition of sulfides. The activation process will convert serine into OAS or O-acetylserine. The serine acetyltransferase or SAT can mediate the transfer of the acetyl group onto acetyl CoA molecules. The O-acetylserine lyase or OAS-TL uses sulfides and OAS as substrates and this enzyme can participate in the beta replacement reactions (Hell et al., 2011). The enzymes OAS-TL and SAT can play important role in the synthesis of cysteine molecules. Most of these enzymes will be accumulated in the cytosol, mitochondria, and plastids. In mitochondria, beta-cyanoalanine synthase was identified. This enzyme can detoxify cyanide that is produced in the process of ethylene production.

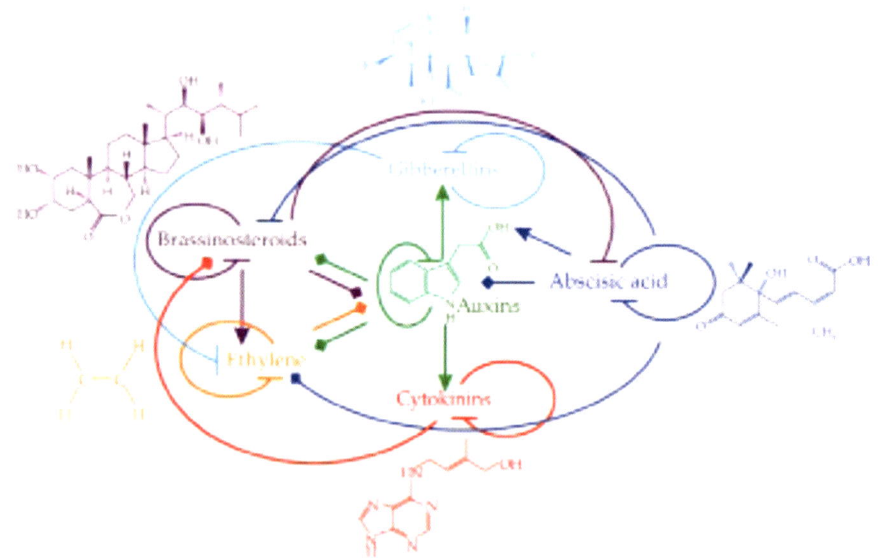

Figure 3.7: Plant hormones.

Source:https://www.nature.com/articles/nsmb0610-642

The plant hormones can participate in the developmental processes. Most of the hormones participate in the nutrient metabolic pathways. The cytokinins are produced from the adenine molecules. The division and differentiation of the cells in the plants can be regulated by cytokinins. The auxins can combine with the cytokinins and regulate the nutrient metabolic pathways. The regulation of SULTR1;1 and 1;2 can be via cytokinin (Maruyama et al., 2004). When cytokinins are added to the growth media, there was decreased sulfate absorption. They can regulate the two transporters at messenger RNA levels. The signaling process in the assimilation of the sulfates is based on the auxins. When sulfate deficiency is detected in the plant system, a molecule called aglycon is produced. This is produced from the IAA or indole acetic acid and indole glucosinolates. IAA can induce root growth. In the case of sulfur deficient plants, the number of lateral roots will be increased and the length of the roots will be extended. There are many studies that can regulate the genes that respond to auxins. The reduced cysteine production can result in OAS deposition in the tissues. When the tryptophan levels increase, there will be the production of auxin compounds. Jasmonic acid or JA can control the sulfate metabolism in the plant cells. Oxidation stress can be regulated by JA. The production of the defense molecules can be regulated by JA. JA can activate the genes that are involved in the production of glucosinolate.

JA can aid in the production of glucosinolates and antioxidants. JA can be used in the production of glucosinolates and antioxidants. These chemicals can be used in detoxification and pathogen defense. These hormones can be useful in maintaining homeostasis in the plant systems. The assimilation of the sulfates can be based on the metabolites in the cellular processes. There is a link between the assimilation of nitrate and sulfates. The nitrogen compounds can enhance the flux of the sulfates. The addition of glutamate and ammonia can increase the influx of sulfates in plants. The carbohydrates can be considered as acceptors of the sulfide molecules. The carbohydrates can work as reductants of sulfates in the tissues that are not photosynthetic in nature.

Chapter 4

Bacteria in Plant Nutrient Management

CONTENTS

4.1 Introduction..72
4.2 Mechanism..74
4.3 Zinc Absorption..82
4.4 Uptake of Phosphate..85
4.5 Sulfur Oxidizing Bacteria..86
4.6 Siderophores..89
4.7 Phytohormones..92

4.1 INTRODUCTION

It is evident that the fuel prices hike enhanced the cost of the agricultural crops and this is the main reason why the demand for alternative energy sources is increasing these days. The crops can be considered as raw materials in the production of alternative energy. It was felt that there will be excess demand for plant nutrients in the coming decades. The developing countries are facing a shortage of plant nutrients. It is important to bring upon many changes in the plant nutrition systems as part of integrated nutrient management systems. Through the integrated nutrient management systems, balance is being established in soil fertility and plant nutrition. The farming systems were upgraded to utilize the fertilizers to the fullest. The use of biofertilizers, minerals and organics have become common in many countries. The economics and social viability were also considered whilst dealing with integrated nutrient management. The increase in plant diseases can be considered as an important factor behind the reduced production of the oilseeds. In the case of intensive cropping, the nutrients are supplied to the crop fields via chemical fertilizers.

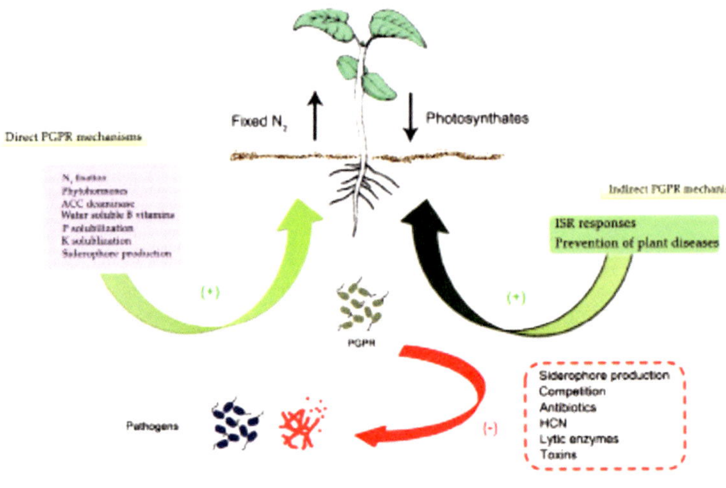

Figure 4.1: Plant growth-promoting rhizobacteria.

Source:https://www.researchgate.net/publication/331876132_Mechanism_of_Plant_Growth_Promoting_Rhizobacteria_PGPR_in_enhancing_plant_growth-A_Review/figures?lo=1&utm_source=google&utm_medium=organic

The effect of the chemical fertilizers will be reduced due to leaching, denitrification, and volatilization. Fertilizers are considered important chemicals in agricultural practices. Increased use of fertilizers has produced several problems to the environment. Biofertilizers were used in place of chemical fertilizers to produce less environmental problems. The biofertilizers provided resistance against several stresses.

One of the important biofertilizers that can be used in the cropping systems are plant growth-promoting rhizobacteria or PGPR. Biopesticides can be used to control the pests of the plant species. The PGPR can be integrated with the chemical fertilizers so that there will be a reduced incidence of the diseases. The organic substances can control the growth of the plants and the oil seeds. They can regulate the morphological and physiological processes. All these processes can get altered even with the application of limited chemicals. Microbes can produce cytokinins, gibberellic acid and indole acetic acid. The fertilizers can be considered as important chemicals in the latest agricultural practices. These fertilizers can supply essential elements needed for normal agricultural growth. The microbial inoculants and biofertilizers can enhance the growth and productivity of the plants. PGPR can enhance crop yields in diverse climatic situations. The inoculation of PGPR can enhance the oil content in the seeds. The agrochemicals can diminish the bacterial growth and most of the beneficial bacterial types can be damaged or eliminated. Different chemicals will have different effects on the microbial populations (Kantachote et al., 2016).

The pesticides were found in the oil cake, oils, and seeds. The crop growth will be reduced and there will be damage to seedlings due to the application of pesticides, weedicides, and fungicides. There are many biocontrol mechanisms that are yet to be revealed. Sesame can be classified as an important oilseed crop for years. The sesame seeds have sesamol, sesamin and sesamolin. There are antioxidants and lesser amounts of saturated fatty acids. Sesame oil can be used to lower the chances of getting cancers. Sesame oil can be used to control cholesterol levels and hypertension. There is an increased demand for sesame seeds. The sesame plants are sensitive to pathogenic fungi. Charcoal rot was considered an important fungal disease of sesame plants. The causal organism for this disease is *Macrophomina phaseolina*. We need to use biofertilizers, inorganic and organic chemicals in order to enhance the productivity of these plants. These chemicals can increase the physical conditions of the soils.

Brassica oil can be considered an important source of oil. Brassica plants are susceptible to several diseases that are caused by viruses, bacteria, fungi, and insects. Damping-off is caused by *Pythium debaryanum* and charcoal rot is caused by *Macrophomina phaseolina*. Fusarium wilt is caused by *Fusarium oxysporium* and stem rot is caused by *Xanthomonas campestris*. There are certain bacterial species that can use ACC. The canola seedlings will have their roots elongated due to the action of PGPR. PGPR can enhance the production of cytokinin and IAA. The groundnut species can be considered as a major producer of oil and it was considered as a food crop. It also comes under the semi-arid tropics. The leaf spot disease is prevalent in groundnut crops. It is caused by the fungus named *Cercosporidium personatum*. This fungal species can incur much loss in the groundnut yields every year. The pod yield will be significant and it enhances the reduction of the pod dimensions and growth. There will be a loss of pods due to leaf spot disease. The rhizobium species can enhance the chlorophyll content of the leaves and so there will be increased plant growth. The rhizobium species assist the plants in the uptake of the minerals. The weight of the nodule, the biomass of the seedling and the germination of the seeds will be drastically enhanced as the incidence of the disease is reduced. There were several positive effects due to integrated nutrient management and PGPR. The phosphorus solubilizing bacteria or PSB can increase the oil yield, oil content and seed number in the sunflower plants.

The utilization of Nitrogen, phosphorus and potassium can enhance the sunflower yields. The use of biofertilizers has improved the growth of sunflower crops. Some of the biofertilizers that can be used to enhance the yield are Phosphobacterium and Azospirillum species. The application of fertilizers can add heavy metals that are toxic in nature. Some of the heavy metals that are prevalent in the soils are cadmium and lead. There are certain fertilizers that can add large amounts of heavy metals to the agricultural soils. The toxic metals can be toxic to humans, animals, and plants. It is important to enhance the fertility of the soils through technologies that are sustainable. It is important to know that the application of PGPR can drastically reduce the application of pesticides and chemical fertilizers. In order to maintain soil fertility, it is important to practice effective nutrient management that involves N, P and K.

4.2 MECHANISM

The excess use of fertilizers and pesticides can enhance the problems linked with human health and the environment. Most of the soil microbes will help in the growth of the plants and these microbes can act as an alternative method to sustain agriculture with limited use of chemical-based fertilizers.

There are many microbial inoculants that are being used by the farmers. Most of the plants are harboring microbial populations and so that the secretions released from the microbes can help in plant growth and development. The bacterial populations that can help the plants are called PGPB or plant growth-promoting bacteria. The plant growth can be increased via biocontrol and stimulation. Due to the secretions from the PGPB, there will be reduced soil-borne diseases. There will be suppression of the pathogens, modified root growth, enhanced tolerance towards abiotic and biotic stress and improved mineral nutrition. Some of the traits that can be from the bacteria are regulated phytopathogenic microbes, regulated ethylene levels, phytohormones, iron sequestration, solubilization of phosphates and nitrogen fixation. Some of the nutrients that can be supplied by these bacteria are iron, nitrogen and phosphates. The important mechanisms that can bring about enhanced nutrient absorption are chelation of iron that is assisted by the siderophores, fixation of nitrogen and phosphate solubilization. The concentration of P in the soils can be about 400 mg/kg. The P molecule can exist in two forms that are classified as insoluble. The organic forms of P can be phosphotriesters, phosphodiesters, phosphomonoesters and inositol phosphate or soil phytate. P can be considered as an important macronutrient. It helps in the growth and development of plants. It is primarily based on bioavailability. P can be supplied to the soils through fertilizers and this may need more energy. In the biogeochemical cycles, the PSB species can participate in the mineralization and solubilization of P. The PGPB can help in the growth of plants. In the solubilization process, there will be the production of citric acid and gluconic acid and these are classified as organic acids with low molecular weight. These organic acids can adhere to the phosphates. The carboxyl and hydroxyl groups can participate in the acidification of the soils. After the necessary reactions in the soil, there will be the production of phosphates that are considered soluble.

The production of protons can be important in the solubilization of the phosphates. There will be the formation of chelating substances. There will be the and production of inorganic acids. Due to the action of bacterial species, there will be mineralization and solubilization of P molecule (Logan et al., 2008). PSB can enhance the P in the plant tissues. The specificity between the soil type and host plant will play a vital role in establishing a connection between the PSB and plants. The soluble P can be easily found in the calcareous soils. The physiological status of the plants can determine the effectiveness of PSB in the utilization of P. PSB can survive in the toxic environment due to cell encapsulation. The alginates will be used to encapsulate the PSB bacterial strains so that the bacteria can survive for longer periods (Bashan et al., 2012).

The effect of PSB can be high if sprayed in the form of cell suspensions. It was observed that the bacterial species when inoculated along with the PSB can enhance the activity in the plants. In some cases, we can use mycorrhizae with PSB. In some cases, we can use non-mycorrhizal fungi with PSB. Iron can be considered as the 4th element on the earth's surface. Iron molecules can be found in the form of oxides, oxyhydroxides, and hydroxides. It is important to know that the organisms absorb limited amounts of iron molecules. The plants and microbes need large amounts of iron molecules. This condition is dominant in the rhizosphere region. In the rhizosphere region, there will be s a competition among fungi, bacteria, and plants (Solano et al., 2013). In plants, the absorption of the iron molecules will occur via two paths. There will be a transformation of ferric iron into ferrous iron.

The dynamics of iron molecules in the rhizosphere will be regulated by the cumulative effect of the microbes, plant, and soil properties. The iron nutrition in the plants can affect the composition of the bacterial species. The transgenic tobacco plants can absorb more iron molecules than wild types. The siderophores can be important in the protection of the plants. The role of the siderophores of microbes was tested using radiolabeled probes. The effect of siderophores in the nutrition of the plants was backed by a lack of symptoms that usually occur when there is iron deficiency. The iron concentration in the roots is high. *Pseudomonas fluorescence* can produce a Fe-pyoverdine complex that can be absorbed by *Arabidopsis* plants. Due to this absorption, there will be an increase in the iron content in the plant tissues. The soil bacteria can increase iron absorption even in heavy metal contamination. The mobility of the metals in the soils can be altered due to the metabolites. The siderophores can adhere to iron, plutonium, zinc, lead, arsenic, nickel, copper, cadmium, gallium, chromium, manganese, and magnesium. The siderophores can reduce the stresses caused due to heavy metal pollution in the soils. PGPB can produce siderophores that can protect tomato, canola, mustard crops. Mutation in the siderophores can offer more protection to the plants grown in the Ni contaminated soils. *S. caprea* can protect the plants from heavy metal pollution in the soils.

It is important to know that agriculture is based on several chemical resources that are produced from nitrogenous compounds. If nitrogen-based chemical fertilizers are used, then there will be several environmental problems and human health concerns. It is important to recognize biological N_2 fixation as an ecologically sustainable substitute that can ensure sustainable agricultural development. It is obvious that nitrogen available in the atmosphere in its native form may not be used by the plants unless converted into ammonia and other biological compounds. Rhizobia can be considered as an important microbial species that can help the plants via biological N_2 fixation. Azospirillum species can be used as microbial or bacterial inoculants. The bacterial species can fix less amount of nitrogen molecules when compared with the absorption of these molecules by many plant species.

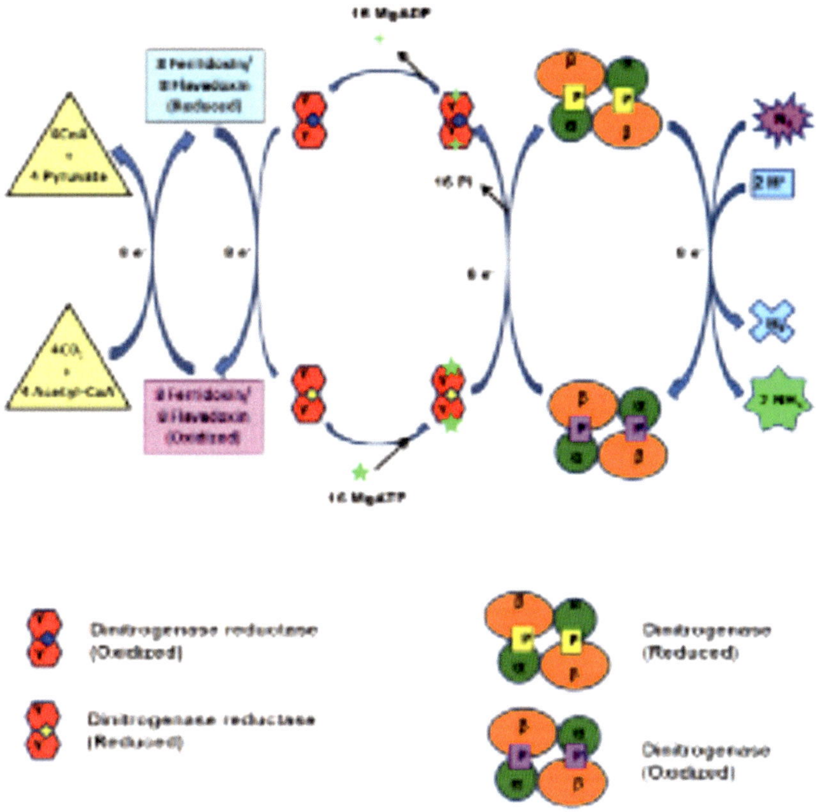

Figure 4.2: Nitrogenase mechanism.

Source:https://www.researchgate.net/publication/321675078_The_Role_of_Microbial_Enzyme_Systems_in_Plant_Growth_Promotion/figures?lo=1&utm_source=google&utm_medium=organic

The nitrogen-fixing bacteria can be called diazotrophic bacteria. This bacterium can fix the nitrogen molecules collected from the atmosphere via the nitrogenase enzyme. It is a metalloenzyme. It contains dinitrogenase reductase and dinitrogenase components. The dinitrogenase reductase can be classified as a dimer. There will be two subunits that are classified as identical in nature. This portion can induce reducing power to the dinitrogenase component. The two subunits can involve in the hydrolysis reactions and will offer binding sites for MgATP. The second component is called as the dinitrogenase component. It harbors a metal cofactor. The biosynthesis of nitrogenase involves nif genes. The nif genes are regulatory genes and structural genes. Some of the genes can induce iron proteins. These genes can aid in the biosynthesis of the Fe-Mo cofactor. They also participate in the donation of the electrons. Most of these genes are necessary for maintaining the function and structural integrity of the nitrogenase enzyme (Figure 4.2). These genes can get integrated with the plasmids and in some cases, they will be encoded by the chromosomes (Fischer, 2018).

The fixation of nitrogen is not an easy process in plant systems. This process demands more ATP. The oxidative phosphorylation reaction in the rhizobium bacteria can be important in the production of ATP molecules. In this process, most of the carbon molecules will be directed to generate ATP but not to produce glycogen. An *R. tropici* strain was produced with the deleted genes for glycogen synthesis. The presence of oxygen can suppress the *nif* gene expression and inhibit the nitrogenase enzyme. In the respiration mechanisms of Rhizobia, oxygen must be essential. The genetic engineering of the bacterial hemoglobin in the Rhizobium species was performed. *R. etli* species can produce three *nifH* gene copies. All these copies will be regulated by a promoter. The reduction of the H^+ to hydrogen molecules via nitrogenase can be considered as an unwanted reaction in the first place. Most of the ATPs will be used in hydrogen production. Half of the electron flux can be transferred to nitrogen molecules. There will be decreased nitrogen fixation.

Rhizobium leguminosarum has eighteen genes that regulate the activities of the hydrogenase enzyme. Eleven *hup* genes are present that are involved in the uptake of hydrogen. The increase in the ethylene levels can reduce the rhizobium population and so the production of the root nodules will be lowered. Some of the strains of rhizobium species can enhance the quantity of the nodules and this can be done by reducing the ethylene concentration. In order to reduce ethylene concentrations, rhizobitoxine can be produced. This chemical can lower the activity of ACC synthase. This intercepts with the biosynthesis of ACC prior to its conversion to ethylene. When the ethylene levels are lowered, the biomass of the plants and the number of the root nodules will be increased. Very limited strains of rhizobium can have an ACC deaminase enzyme.

The abscisic acid or ABA, ethylene, gibberellins, cytokinins and auxins can play important roles in the growth and development of the plant species. In the case of extreme environmental conditions, the concentration of phytohormones in the plant systems may not be adequate to evoke a response so that plants can sustain their normal growth. There are certain phytohormones that can activate the tuber growth, seed germination, fruit ripening and formation of roots. Most of the microbes that inhabit the rhizosphere can alter the course of phytohormones. The hormonal balance in the plant systems will be affected by PGPB. IAA can be considered as an important phytohormone that can participate in all the growth stages of plant life. This hormone can aid in cell differentiation, extension, and division. They can even intervene in the germination process of the tubers and seed germination. They can enhance the development of roots and xylem tissues. Most of the processes that are necessary for the vegetative growth of the plants can be managed by auxins. These hormones can help the plants in evoking immune responses and they aid in stressful conditions. The fructification process, formation of essential pigments and the rate of photosynthesis can be affected by auxins. The responses towards gravity and light can be due to auxin levels in the plant systems. Indole-3-acetic acid can be considered as an important type of auxin that has been studied in detail.

The auxin concentrations can affect the physiology of the plants in many ways. The effectiveness of the auxins will differ with the sensitivity of the plant tissues and the concentration of the auxins applied. The auxins that are present endogenously are sensitive to the microbes in the soils. These plant-derived auxins can activate the bacterial auxins. The bacterial auxins can affect the plant growth positively and negatively in some cases. The IAA molecules that are produced endogenously can help the plants in their optimal growth. The auxin production is prevalent in the bacteria present in the soils. The tendency of producing the auxins can be observed in archaea, cyanobacteria, methylobacteria and streptomycetes. The biosynthesis of auxins will be based on tryptophan levels. Tryptophan can be considered as a precursor for IAA production. The pathways that can yield IAA can be classified into expressed pathways and inducible pathways.

The biosynthesis of auxins in bacterial species can depend on many factors. They are root exudates, starvation of carbon molecules, stress on matrix, osmotic stress, pH, and environmental stresses. The main effect of IAA produced by the bacterial species can be the production of adventitious roots and lateral roots. Due to the exaggerated growth of these roots, there will be an enhanced uptake of nutrients and minerals. The root exudates can activate the proliferation of the bacterial species ,especially in the roots. Many members of the rhizobium species can produce IAA. It was observed that the auxin levels can aid in the organogenesis of the nodules. The IAA concentration in the nodulated roots was high. The PGPB species can produce IAA molecules and they aid in the avoidance of environmental stresses. IAA can help in the elongation of shoots and roots in saline conditions. IAA can be biodegradable in nature. The bacterial species can catabolize the IAA molecules. This can be observed in the plant tissues also.

Cytokinins can be called aminopurines with N6 substitution. They have a major role in several physiological processes. They intervene in the delaying of senescence, expansion of leaf, chlorophyll accumulation, growth of roots, seed germination and cell division. These chemicals can control the gene expression of expansin protein. This protein can loosen the cell walls of the plants. This will help in the cell expansion that is basically turgor-dependent. This can control the size and shape of the plant cells. The cytokinin production was first observed in the Agrobacterium species. They were later observed in methanotrophic and methylotrophic bacteria. The genes that can encode IPTS or isopentenyl transferase was first identified in 2001. This observation has proved that the plants can produce isopentenyl transferase that was classified as an enzyme needed for the biosynthesis of cytokinin. The cytokinin producing bacteria when inoculated on seeds can produce high cytokinin content. This can enhance the growth and development of plants. There are many environmental stresses that can increase cytokinin levels. The ethylene levels in the plant species will be increased resulting in diminished root elongation. The tendency of rhizobia to produce nodules and the level of cytokinins produced can be correlated. The cytokinin can help in the differentiation of nodules and rhizobial infection. *Bacillus megaterium* can stimulate the plants to grow branched lateral roots and shorter tap roots. Earlier studies indicated that these effects were due to cytokinin activity.

Gibberellins can be produced by bacteria, fungi, and plants. They are characterized as diterpenoid acids. They have isoprene residues and mostly we can observe four rings. There are about 136 gibberellins (MacMillan, 2002). These chemicals can intervene with cell elongation and cell division.

The delaying of senescence, setting of fruits, flowering, elongation of stems, germination of seeds and other developmental processes can be regulated by gibberellins. This phytohormone can enhance the root growth and the abundance of the root hairs can be observed on the application of gibberellins. The hormonal balance of the plant species can be altered by the action of gibberellins. *Azospirillum brasilense* synthesize gibberellins like analogs. In the roots of a plant, the IAA molecules will be present endogenously and this will help in the optimal growth of the plants. Auxin production is common in most bacterial species. The concentration of gibberellins can be altered due to chemical activation and deconjugation. The production rate of gibberellins can be an important factor in determining the concentration levels in the plants. The deconjugation process will occur in the reactions involving glucosyl gibberellins. The inactive gibberellins can be activated due to the action of the PGPB species. The Azospirillum species can fix the atmospheric nitrogen. It can produce IAA molecules and so be involved in the plant growth and development. The synthesis of the gibberellins can be linked with the activity of *Azospirillum* species.

Ethylene can be produced in higher plants with the aid of 3 enzymes. They are ACC oxidase, 1-aminocyclopropane-1-carboxylic acid or ACC synthase (Kende, 1989) and S-adenosyl-L-methionine synthetase or SAM synthetase. The SAM synthetase can convert methionine into SAM. ACC synthase can catalyze the SAM hydrolysis resulting in the formation of 5'methylthioadenosine and ACC. The enzyme ACC oxidase can convert ACC into cyanide, carbon dioxide and ethylene. There are many steps in the growth of the plants. They are abscission of petals, leaves, senescence of flowers and ripening of fruits. All these steps are controlled by ethylene. This hormone can control plant metabolism whilst abiotic and biotic stresses (Kang et al., 2010). The production of ethylene will be high if there are environmental stresses. The stress can be due to fungi, bacteria, viruses, high salt concentration, predation by insects, wounding, radiation, organic pollutants, toxic metals, drought, flooding, excess light, and temperature extremes. The production of ethylene can be based on the nature of stress and the age of the plant. The plants can be exposed to several stresses and so there will be complete protection to the plants. The genes linked with pathogenesis can get activated so that the resistance in the plants will be activated. The PGPB strains can get populated in the roots and seeds of the plants. This can be due to tryptophan production. Most of the bacterial species can produce IAA. The bacterial colonization in the root zones can produce ACC deaminase. Due to this, the ethylene production will be limited. The nodule formation and the association of VAM can produce ethylene content. The bacterial species that can produce ACC deaminase can reduce the production of ethylene resulting in colonization with VAM and nodulation. Bacterial species can encourage the growth of the plants and due to the production of ACC deaminase, there will be a high rate of plant growth and this was evident during the flooding seasons (Grichko et al., 2001).

The suppression of plant diseases can be due to soil microbes and they can lower the utility of chemical fertilizers. The biocontrol of the pathogens can be based on several factors. It is based on systemic resistance, enzymes, antibiotics, colonization, and competition. The colonization in the root systems can be competitive in nature and this will help in the biocontrol process. The PGPB species can control the pathogens by producing organic compounds so that the microbes that cause diseases will be destroyed. There are several factors that can affect the role of PGPB in the growth of the plant tissues. The plasmids, root exudates, relative humidity, temperature, and composition of the soils can be classified as important factors in PGPB based plant growth. The production of antibiotics can be linked with the tendency of the PGPB strains to reduce the development of pathogens. Some of the antibiotics that are produced by PGPB are pyrrolnitrin, pyoluteorin, phenazines, hydrogen cyanide, oomycin and others. Some of the enzymes that help in the biocontrol process are lipase, protease and chitinase. Chitinase can control the fungal pathogens. It can control *Pythium ultimum*. Chitinase can regulate *Fusarium* species. PGPB can act as a biocontrol agent and this was possible only through enzymes. The important enzymes are lipase, protease and chitinase.

4.3 ZINC ABSORPTION

Zinc can be used in several living entities and high Zn concentrations can lead to toxicity. The toxicity due to zinc in the soils is rare. Mostly the living systems will be deficient in zinc. The zinc deficiency in humans can lead to several health problems like abnormal brain development, immunity disorders and growth abnormalities. (Figure 4.3)

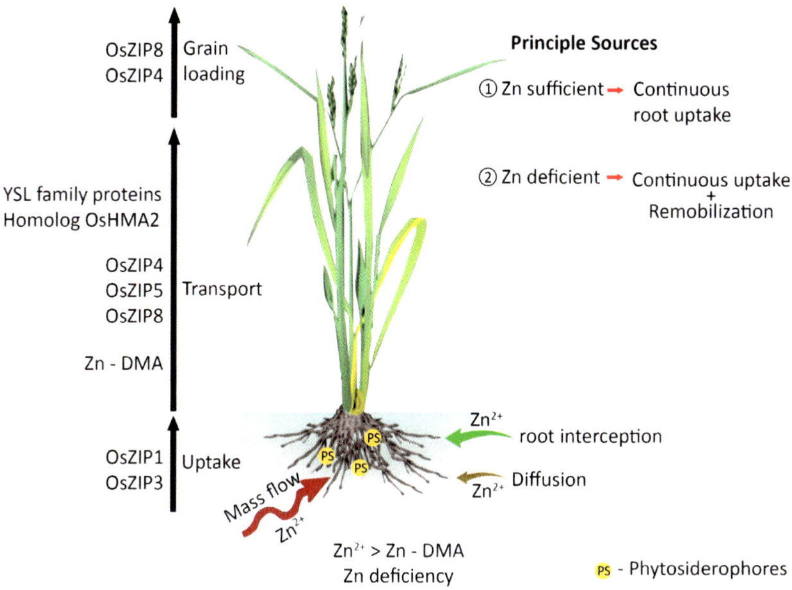

Figure 4.3: Zinc absorption.

Source:https://www.frontiersin.org/articles/10.3389/fpls.2016.00764/full

It is important to supply zinc supplements to the crops and ensure crop rotation regularly. Most of the soils are deficient in zinc and it was reported that most of the soils face the problem of zinc deficiency. Bacteria can solubilize the zinc compounds and so the availability of Zn will be high in the soil solution (Saravanan et al., 2007). Zinc compounds can exist as cations. The Zn^{2+} ions can participate in the living systems. The size of zinc atoms can be the same as other cations like copper, nickel, cobalt, and iron. The plant cells may not have the capacity to resolve metals that are useful or toxic in the process of absorption. Most of the metals that are cations can be transported into the cell membrane. This can be done via chemiosmosis. In this process, there will be metal inorganic transport or MIT. The transporters are not specific in nature and they are classified as open-gate systems. This can result in metal toxicity in the plants. Zinc participates in the functioning of 300 metalloenzymes. This compound can be present in ligases, isomerases, lyases, hydrolases, transferases, and oxidoreductases.

The role of zinc in the transcription process is important. Zinc molecules can build a domain that can be detected by lipids, proteins, and nucleic acids. It is present in the DNA binding proteins. TFIIIA can be considered as a transcription factor that has zinc. According to proteomics, at least 10% of proteins have binding proteins with zinc. Raulin in 1869 identified zinc molecules in the living systems like bread mold. Zinc can be present in the soluble forms in the plants. Zn participates in reproduction, auxin metabolism, the integrity of membranes, protein metabolism, the formation of starch and sucrose, photosynthesis, and the metabolism of carbohydrates.

Zinc has an important role in the evolution process. In the evolutionary process, most of the biological elements are affected by zinc. Cyanobacteria has maintained its integrity with the radiation in the process of evolution. The concentration of zinc will be based on the crop type and the geographical location. It is important to know that, the zinc concentration should not fall below 0.5mg/kg of soil. The symptoms associated with zinc toxicity are common in microbes and plants. The zinc pollution is evident due to the dumping of zinc wastes from the mines. The flood events in the alluvial soils can result in the accumulation of zinc. The sewage sludge may have excess zinc content. Most of the plants can tolerate excess zinc levels. Chlorosis occurs in the case of zinc toxicity.

The soils can be declared as toxic with zinc only if the concentration is between 100 and 500mg/kg of soil (Macnicol et al., 1985). Due to abnormal Zn levels, microbial diversity will be affected. The diazotrophs can be more sensitive to zinc toxicity. *Bradyrhizobium* and *Rhizobium* can be considered as sensitive species as far as the Zn toxicity is concerned. Zinc molecules can enter the cells via non-specific and specific transport systems. The electron transport chain in the respiration process will be inhibited due to toxic Zn concentrations. The microorganisms have developed many mechanisms that can help in gaining resistance against Zn toxicity. Some microbes can detoxify excess zinc. The excess zinc compounds can get associated with organic acids. The uptake of zinc by the plants can occur via metabolism-based process and energy independent process. In the case of bacteria, the uptake of zinc will be regulated by plasmids.

There are many kinds of zinc resistance mechanisms in bacterial species. The plasmids present in the bacterial cell can induce resistance against the Zn toxicity. The RND (Resistance Nodulation Division) exporters belong to the protein family that helps in resistance, nodulation, and division (Tseng et al., 2013). Zinc element can be present in the soils in diverse forms. Some of the forms of Zn are zinc sulfide, willemite, zinc silicates, sphalerite, smithsonite and zincite. The solubility of Zn can get increased if pH levels are reduced. There are many bacterial species that are present in the rhizosphere region.

These microbes can affect the availability of zinc in the plants. *Abisidia cylindrospora* and *A. spinosa* can solubilize zinc phosphate and zinc oxide (Gaither et al., 2001). The addition of ZnO can induce buffer action. The mycorrhizae can exhibit two important mechanisms and it depends on the availability of the Zn element. The solubilization process can be increased by the addition of oxalic acid. VAM can mobilize the available Zn and can act as a solubilizer. If the concentration of Zn is high in the environment, there will be an abnormal Zn accumulation in the tissues.

4.4 UPTAKE OF PHOSPHATE

The fixation of phosphate can be considered as an important problem in agricultural soils. The fixation of phosphate ensures that the phosphates are removed from the soil solution and transported into the soil solid phase. The precipitation of phosphates can occur due to ferrous and aluminum ions that are present in the soil solution. The sorption of phosphate in the soil minerals can be important. The sorption of phosphates can occur via specific sorption or non-specific sorption. The phosphate ions can be exchanged with -OH groups in case of specific sorption. In the case of non-specific sorption, there will be an electrostatic attraction of the ions on the soil surface. Histosols can encourage phosphate sorption at high rates. The ascending sequence of the sorption of the phosphates are andisols, ultisols, oxisols, mollisols, vertisols and histosols. We can quantify the concentration of soluble phosphates via the Mo-Blue method. There are several soil microbes that can solubilize the phosphates. Aspergillus and Penicillium species can be classified as phosphate solubilizers. The acidic conditions can be important in the solubilization process.Gluconic acid released by the microbes can be vital in the solubilization of the phosphates. These acids are primarily produced in the processes of microbial metabolism like fermentation and oxidative respiration.

The solubilization of the phosphates can occur through several reactions in the rhizosphere area. Bacterial species can be important in the solubilization of the rock phosphates. The solubilization of the microbes can be linked with the presence of the Ca phosphates. The tricalcium phosphates can be important whilst studying the solubilization of the phosphates (Osorio et al., 2012). The solubilization of the calcium phosphates is high. The solubilization of the Fe phosphates and Al phosphates can be studied in lab conditions. The Fe and Al ions can get chelated with organic anions like citrate and oxalate. This will dissolve the Fe and Al phosphates. Phosphate solubilizing microbes can release several organic acids that can solubilize the phosphates from Fe and Al phosphates and this is possible in acidic conditions. Most of the research is focused on solubilization due to microbes. The addition of the phosphate solubilizing microbes can exhibit a positive impact on plant growth and yield. Excess carbonaceous substances can be released into the rhizosphere region by the VAM species. The addition of Rhizobium, mycorrhizal fungi and PSM can enhance the uptake of the phosphates. The pseudomonas species and rhizobium species can induce the nodulation process in the legumes. Application of phosphate solubilizing microbes can enhance the phosphate sorption in the less weathered soils. The effectiveness of the PSM species was analyzed by Osorio and Habte in 2001.

4.5 SULFUR OXIDIZING BACTERIA

Sulfur can be considered an important nutrient in plant metabolism. S is present in amino acids like methionine and cysteine. These amino acids can be considered as important as they involve in many reactions occurring in the living systems. The Sulfur element can be important in animal and plant life. A high sulfur deficiency was observed in Oxisols, ultisols, Alfisols, Vertisols, and Andosols. The S deficiency can be reduced by using the fertilizers enriched with S. The S deficiency in the soils can be due to reduced rate of atmospheric deposition, intensive cropping, reduced use of organic manures, and usage of low S-based fertilizers. In order to reduce the S deficiency, we need to go for S oxidizers that can enhance the formation of the sulfates. The S molecules are present in the soils in the form of organic S compounds, hydrogen sulfides, sulfates, mineral sulfides, and elemental sulfur. Most of the S compounds will be from the pyrites or FeS_2. In the formation of the soils, the elemental S can exist in many states. The redox reactions of sulfur can occur instantly. There are diverse S reactions. The S present in the soils can have interesting properties. They are soil texture, aluminum oxides and active iron, the percentage of clay and organic matter present in the soils.

The S present in the organic form that is in the soils can get enhanced due to the presence of the humus. The marsh and peat soils can be rich in the S molecules that are organic in nature. The total S present in the soils can have less organic matter. The tropical soils that are leached and weathered can have less organic matter and so the S content will be less. The S content can be directly proportional to the clay content of the soils. The estimation of total S may not be enough to know the exact S composition. The organic sulfur can be classified into 3 portions. They are soil biomass, on-C bonded and C bonded (Freney, 2015). The C-bonded S can be present in the semi-arid regions and humid regions. Methionine, and cysteine can be important amino acids that are having S-elements in their skeleton. Most of the S will be in the form of sulfolipids. The S compounds present in the microbes can be important as this determines the availability of S to the plants. The S molecules will be present in the form of sulfates and the plants absorb sulfates from the soil solution. The sorption of the S compounds can be due to ion exchange in the clay minerals with a decrease in pH value. Pentathionate can exist in -2 and +6 states. Tetrathionate and trithionate can exist in -2 and +6 oxidation states. Dithionate can be +4 oxidation state and hyposulfite can be in a +3 oxidation state. Sulfuric acid and sulfate can exist in +6 oxidation states. The sulfites can exist in a +4 oxidation state. The thiosulfates can exist in -1 and +5 states. The elemental sulfur can be in 0 oxidation state. Pyrites can occur in -2 and 0 states. The hydrogen sulfide and sulfide can exist in the -2 oxidation state. The sulfates can be considered an important source for S in crop plants. APS or adenosine-5'-phosphosulfate can be important in the activation of the sulfates. The sulfolipids can be important in the photosynthesis process and electron transport system. The sulfur molecules can be important in the production of lipids. There are many sulfur-containing compounds that are present in living organisms. Some of them are chlorosulfolipids, phosphatidylsulfocholine, sterol sulfur ester and ceramide sulphonic acid. The carotenoids and flavonoids can be produced with the aid of sulfur-based compounds. The polyamines can be considered as important growth regulators that can help in the hormone-mediated processes. The glucosinolates can be classified as non-volatile substances. These compounds can be detected by the insects only when comes into contact. The isothiocyanates can be classified as volatile and the insects can respond to the scents. The isothiocyanates can have antifungal and antibacterial properties. The Brassica species can limit the infection of the mildews. The glycine betaine and proline can be important compounds that can be used to combat the saline conditions. The reduction in the S molecules can also lower the Mg, Ca, Fe, Si and N molecules (Martin et al., 2008). Allyl isothiocyanate can be an important condiment used by humans. Isothiocyanate can be present in the crucifers and cabbage (Plonka et al, 2011). Due to the application of sulfur-based compounds, the proportion of the allylthiocyanate in the seeds can be increased. The legume plants need sulfur and phosphorous compounds. If the S and P composition in the soil is reduced, then the growth of the crop will be hindered. The application of S-based fertilizers can enhance shoot length, dry matter, the biomass of the nodules and pod yield. The oil content will be increased. More glucosides will be hydrolyzed on the application of S. In the case of mustard, the oil content and seed yields were enhanced. Enhanced plant growth was observed in sunflower, sorghum, cotton, wheat, maize, and rice plants.

The sulfur molecules can be present in the enzyme nitrate reductase. This enzyme is important in the conversion of nitrates. Organic nitrogen will be produced. Due to S deficiency, the nitrogen metabolism will be hindered. Due to reduced S content in the plants, there will be light green coloring. It is important to know that the deficiency symptoms will be observed in the alfalfa plants as they need more S content. The availability of the anions in the soil solution can be based on other cations like potassium, sodium, magnesium, and calcium. The sulfate availability will be based on other cations that are present in the soil solution. The cations are important in maintaining a balance between the sulfates and the soil solution. Potassium sulfate, potassium magnesium sulfate, superphosphate, magnesium sulfate, ammonium thiosulfate and ammonium sulfate can be classified as highly soluble. The calcium sulfate molecules can be classified as least soluble and the elemental sulfur is insoluble in nature.

The oxidation of S can be based on the nutrient availability and redox potential of the soils. The sulfur molecules have a synergistic effect (Tiwari, 2012). The microbes can participate in the biogeochemical cycles of the sulfur molecules. There are many organisms like green sulfur bacteria and prokaryotes that can use the sulfur compounds. These molecules can be considered as electron donors. There will be oxidation of the sulfates. The prokaryotes can oxidize the sulfur molecules effectively. Some of the microbes can generate energy through metabolic activities and this can be due to the oxidation of the compounds that are inorganic in nature. Some of the carbon compounds like carbonates, bicarbonates and others can get assimilated. If the sulfur concentrations are lowered, there will be limited growth. It is important to ensure that there should be co-oxidation of the sulfur compounds along with the substrates that are organic in nature. This ensures plant survival and growth will be excellent.

Bacterial species can be important in the oxidation of the thiosulfates. *Thiobacillus thioparus* was isolated from the salt water and fresh water bodies (Boden et al., 2017). This species can be found in the seawater, sewage, ditchwater, and soils. This species is colorless and they may not form threads. The sulfur molecules can be used as an energy source. The S oxidizing bacteria were isolated from the clay loams. The sulfur-oxidizing bacteria are rarely found in the alkali soils. *S.novella* and *A.thiooxidans* were isolated from the alkali soils. The S oxidizing bacteria were isolated from the natural habitats. The differential media with varying pH can be used to resolve the acidophilic and neutrophilic species. The probes made up of oligonucleotides were devised to detect the *Acidiphilium* and *Thiobacillus* species from the samples collected from the environment. The sulfur bacteria, phosphobacteria and Thiobacillus can be applied to enhance the maize growth (Balakrishnan et al., 2017). The sulfur-oxidizing bacteria can be used to enhance the dry weight of the pods, leaf width, length of shoots and roots. The sulfur-oxidizing bacteria can reduce the growth of *Leptosphaeria* and *Rhizoctonia* species.

4.6 SIDEROPHORES

The soils can be classified as oligotrophic environments in which there will be the growth of microbes even in limited conditions. Most of the microbes must fight for the resources that are limited. It is important to understand the diversification of the microbial needs so that the purpose of the elements in crop production can be formulated. The sulfur-oxidizing bacteria can reduce the growth of several pathogens. Iron can be found in the crust of the earth. The iron molecules can be compared with copper, silver, and gold. Iron can be considered as the 4th abundant element in the crust. The iron content present in the soils can be present in the silicates or hydroxides or iron oxides. They can be used by plants and microbes.

The iron molecules that can be assimilated can be called as bioavailable forms. Most of the iron molecules are present in the form of an aqueous solution. Iron molecules can exist in two oxidation states. They are ferrous state or Fe^{2+} and Ferric or Fe^{3+} state. Iron can be classified as an essential nutrient in plant growth and development. The need for iron inliving organisms varies (Raymond et al., 2003). The plants require 10^{-9} M of iron for optimal growth. Iron deficient soils are found everywhere as the plants demand more iron. Iron and Iron-sulfur clusters can work as catalytic centers for several redox reactions. The role of iron is in the photosynthesis, DNA synthesis, synthesis of nucleotides, amino acids, activation of oxygen and in electron transport mechanisms. The ferrous substances that are classified as insoluble may not be available for plants. In the neutral pH conditions and aerobic conditions, the insoluble ferrous molecules are prevalent. In the case of anaerobic conditions, the ferrous molecules are prevalent. Gram-negative bacteria have porins that can absorb the iron molecules that are in soluble forms. The cytoplasmic proteins that can store the iron molecules can be called as ferritins. These proteins can get activated to supply iron molecules in case of any shortage. These proteins can protect the cells against iron toxicity.

Heme can be considered as a prosthetic group that is associated with several enzymes. Due to its toxicity, the heme group is always bound and rarely free. Heme can be classified as an iron source for many bacterial species. The iron source for many bacterial species can be hepatoglobin-hemoglobin complex. There are many types of hemoproteins. The heme proteins may not be linked with the apoproteins. The hemophores are present in the gram-negative bacteria and they are classified as extracellular proteins. The microbes have developed several mechanisms that can help them to adapt to the iron deficient conditions in the environment.

Siderophores can be classified as iron specific chelators that have low molecular weight. They can be considered as iron carriers. These molecules can help in the biosynthesis of ferric iron (Dhungana et al., 2005). The siderophores were extracted from *Ustilago sphaerogena*. There are several organisms that exhibit different types of siderophores. These siderophores can be classified based on functional groups. The siderophores can chelate the iron in its ferric state. The siderophores can be classified into 4 types and this is based on the functional groups they carry. The hydroxamate type of siderophores can have hydroxamic acids. The acids found as the functional groups are rhodotorulic acid, ferrichrome and anguibactin. The catechol rings have hydroxyl groups that can associate with the iron molecules. In the case of citrate type of siderophores, the carboxylate ions can assist in the binding of iron molecules. *Bradyrhizobium japonicum* can produce this type of siderophores. In complex siderophores, there will be different iron-chelating functional groups that can be combined. Anguibactin and mycobactin can be considered as complex siderophores. There are many Pseudomonas species that can produce siderophores. Pseudobactin and pyoverdines can be produced by this species. The phytosiderophores can be produced by several graminaceous plants. The roots of these plants can secrete these siderophores in order to absorb the ferrous ions. Mugineic acid or MA can be an example of phytosiderophores.

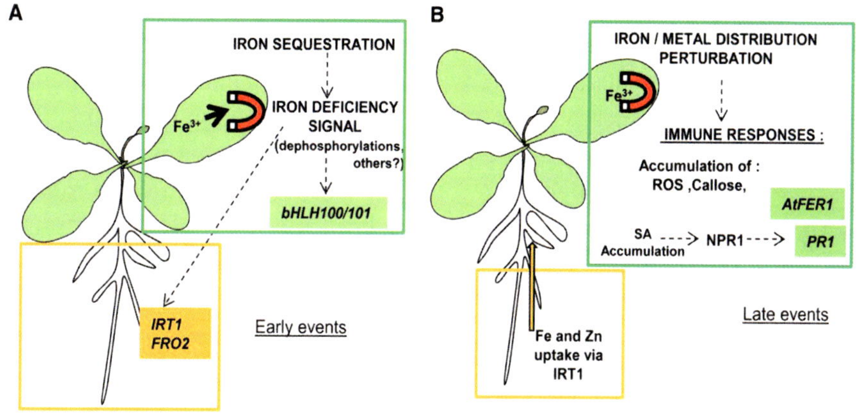

Figure 4.4: Siderophores.

Source:http://www.plantphysiol.org/content/164/4/2167

MA can be associated with biochemical precursors. The siderophores (Figure 4.4) can be based on the competition between the microbes. Iron can react differently in diverse pH conditions and aquatic medium. There is always less iron content in the soils than that needed by the plants. This is even worse in the calcareous soils. The plant species have adapted to the Fe limiting environments. The production of organic acids and proton extrusion can enhance the acidification of the soils. The siderophores secreted from the microbial populations can exhibit a high range of affinity towards Fe chelation or Fe (III) chelation. Due to the siderophores, there will be increased bioavailability of the iron molecules in the plants. The siderophores can have low molecular weight and can exhibit high affinity rates for ferric ions as chelators. This can occur due to iron deficiency. The utilization of these siderophores can be classified as a receptor dependent process. The ferri-siderophores can be multifunctional in nature. The affinity of these siderophores can be based on the composition of the amino acids and structure. The chelator produced by one bacterium can be used by the other bacteria. The fungal species can produce ferrichrome and this can be used by *E. coli* to extract the iron molecules. The derivatives of the ferrichrome can be used as the primary source of iron by the plants. The nitrogen fixation can be done via rhizobia and it needs several proteins that are enriched with the proteins. Some of them are cytochromes and leg-hemoglobin. The siderophore receptors are diverse and there are several heterologous siderophores. The absorption of the iron molecules and the nitrogen fixation is related but not yet established by the researchers. IROMPs or Iron regulated

Outer Membrane Proteins are present in rhizobia and they are involved in both production and secretion of the siderophores. The uptake systems are having high affinity rates and there will be competition between the microbes. These systems can increase the survival rates of living organisms. There will be competition between the pseudomonas in the rhizosphere regions. The rhizosphere in the plants can be classified as a dynamic environment and there will be extreme competition within the microbial communities. The heterologous and homologous siderophores can help in the nitrogen fixation. The siderophores that are classified as hydroxamate can be found in the soils. There are several side-chain reactions that can occur in this process. The peroxides and oxygen radicals can be considered destructive in nature. They can damage nucleic acids, proteins, and lipids in the cell environment. There are many types of uptake regulatory systems. These systems are mostly present in bacterial species.

4.7 PHYTOHORMONES

The soils can be considered as important support platforms for plant and microbial growth. Some of the microbes that can sustain in the soils are fungi and bacteria. These species can proliferate easily. The microbes can grow well in the root systems and that portion can be called as the rhizosphere (Luna et al., 2007). The bacteria that are present in the rhizosphere region can be called as rhizobacteria. The rhizobacteria can colonize the exterior and interior surface of the roots. These species can establish a symbiotic relationship with plants. Some of the species that belong to rhizobacteria can survive as endophytic bacteria. The rhizosphere bacteria belong to Azospirillum. This species can increase plant growth.

This is possible with the production of phytohormones. The important phytohormones are cytokinins and IAA auxins. These bacterial species can also produce ABA, ethylene, and gibberellins. Auxins can be important chemical compounds that can help in the elongation of the cells. The growth patterns are observed in the subapical regions. IAA can inflict physiological effects on the reproductive systems. These chemicals can affect phototropism and gravitropism. The differentiation of the vascular tissue can be affected by IAA. The elongation of roots and stems will be affected by the concentration levels of the auxins. The initiation of the roots and apical dominance can be due to the presence of auxins. The production and metabolism of the auxins in the bacterial species can be important so that they can survive in rough environmental conditions. The biosynthesis of IAA in the plants can be due to Indole-3-pyruvate pathway. This pathway can be called as IPyA pathway. Enterobacter, Rhizobium, Azospirillum and Bradyrhizobium can follow this pathway. In this pathway, Trp is converted into IPyA. IAM or Indole-3acetamide pathway can be present in the bacteria. In this case, IAM is first converted into trp-2-monooxygenase. This enzyme is produced by the *iaaM* gene. IAM can be transformed into IAA. IAM hydrolase can mediate this process. IAM can be classified as an endogenous metabolite that can be isolated from the *Arabidopsis* species (Pollmann et al., 2003). The Tryptamine pathway is followed by *Bacillus cereus*. The IAA molecules can be associated with amino acids and sugars. These chemicals are stored, transported, and protected from the catabolic enzymes. The conjugates can regulate the IAA concentrations in the cells and this can lead to homeostasis. The biosynthesis of IAA in the prokaryotes can take two primary pathways. They are IPyA pathway or indole 3-pyruvic acid pathway and indole 3-acetamide pathway or IAM pathway. There are four pathways that can be considered as important in the Azospirillum species.

The biosynthesis of IAA (Figure 4.5) can be controlled by several factors. The IAA biosynthesis can be linked with plant signaling and environmental stress. The limitation of the carbon source, matrix stress, osmotic stress and acidification can be considered as important factors in the biosynthesis of the IAA molecules. The chemical signals produced by the plants can be considered as secondary factors that can control the IAA biosynthesis. In acidic pH and limitation of carbon molecules, there will be increased production of IAA (Broek et al., 2005). The reduction of the growth rate can a limitation of carbon can be linked with the physiology of bacteria. The production of IAA can be observed in every stage of the bacterial culture and it is important to know that the stationary phase of bacteria can produce more IAA. The exogenous auxins can be produced in the higher plants. These molecules can be produced by bacterial species in the rhizosphere region. Most of the bacterial species can produce compounds that resemble IAA in lab conditions. The mechanisms involved in the production of the IAA compounds can involve translocation of xylem and biosynthesis of conjugates. There will be a continuous interaction between the rhizosphere and the chemicals released from the plants like nucleotides, vitamins, sugars, and other organic acids.

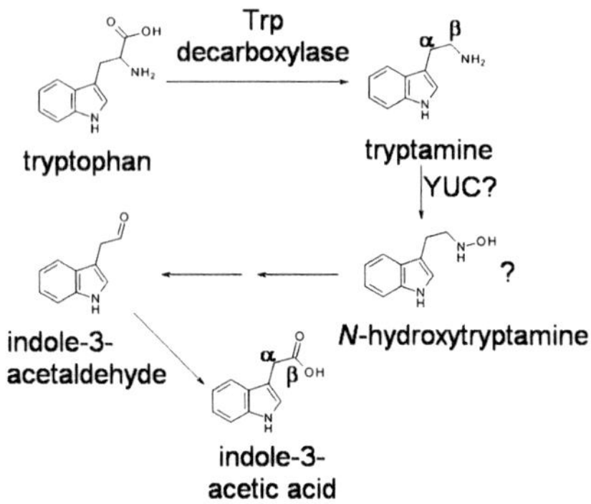

Figure 4.5: Biosynthesis of IAA.

Source:http://www.plantphysiol.org/content/154/4/1957/F1

Gibberellins can be classified as tetracyclic diterpene acids. They can control the growth and development of the plants. Events like fruiting, flowering, elongation of stems and germination can be regulated by the gibberellins. Most of the gibberellins are subjected to carboxylation and this can occur at the C_7 position. The hydroxylation process can be considered as important in the biological systems. The glucose molecules will be adhered to the gibberellins. There will be glycosidic esters and hydroxyl groups. The glucose molecules can bind with the hormones. Most of the conjugated molecules can bind with the glucose molecules. The production of the gibberellic acid in the higher plants can occur with the aid of a 20 carbon molecule compound. It is called GGPP or geranylgeranyl pyrophosphate. It is produced in the plastids. There are approximately 130 GAs (Crozier et al., 1985). Most of the GAs are present in plants and 15 GAs are present in fungal members. Less number of GAs are active in the biological systems (Hedden et al., 2000).

Cytokinins can be classified as natural compounds that can control the differentiation of the cells and division of cells. They mainly act on the meristematic tissues in the plant systems. Mostly, these chemicals are classified as purines and they are produced from adenine. The hormones in the plants can be linked with both cellular and physiological processes. Some of the processes that can be regulated by the cytokinins are leaf expansion, initiation of the stem, elongation of roots, the formation of root hairs, development of roots and formation of organs. Kinetin was the first cytokinin discovered. This chemical was classified as a phytohormone (unnatural). Zeatin was isolated and then classified as a natural form of phytohormones. There are several microbes that inhabit the rhizosphere. These microbes can produce cytokinins in the artificial medium. It was reported that at least 90% of the microbes present in the rhizosphere regions were able to produce cytokinins or their analogs (Barea et al., 1976). PGPR can produce cytokinins.

Ethylene can be utilized in the growth and development of plants. The ethylene molecules exist as gases, so their detection was problematic in physiological conditions. There are several reports that support the production of hormones in plants. Ethylene molecules can be classified as symmetrical. There will be four hydrogen atoms and two carbon atoms in this molecule. This gas can be soluble in water. This gas can induce several physiological changes in the plant tissues even applied in minimal concentrations. In the plant systems, almost all the cells can produce ethylene gas and this can be linked with the developmental phase of the plant systems. If ethylene gas is in higher amounts, we can say that there is an active cell division. This can happen in stressed conditions. The senescence stage can be detected by excess production of ethylene. Ethylene can be produced by *Bacillus* and *Azotobacter* species in the culture medium.

Abscisic acid can be present in many developmental stages. The ABA can take part in the ripening of the fruits, seed dormancy and bud formation. ABA can also participate in controlling abiotic stress conditions. The abscisic acid is produced in the plastids of roots and leaves. The biosynthesis of abscisic acid can be from MA. In this process, Farnesy farnesyl pyrophosphate or FFPP will be produced and this occurs in the biosynthesis of terpenes. The ABA molecules can regulate the water budget in the plants and so it affects the stomatal movement. ABA production in excess amounts can be treated as a stress signal in plant studies. Polyamines can be classified as regulatory compounds. The polyamines can be classified as organic compounds with low molecular weight. They have one or more amino groups. Cadaverine can be used by the plants to control root growth.

Chapter 5

Molecular Basis of Nutrient Efficiency

CONTENTS

5.1 Introduction 98
5.2 Root Systems 98
5.3 Rhizosphere 103
5.4 Canopy Physiology 108
5.5 Remobilization of the Nutrients 109
5.6 Grain Composition 111
5.7 Genetic Improvement 112
5.8 Molecular Genetics 113

5.1 INTRODUCTION

The molecular basis of the nutrient use efficiency is an important in factor that involve genetic improvement. The plant breeding methods can produce high yielding varieties and these crops need more nutrients. Climate change and land use can negatively impact agriculture. Excess utilization of non-renewable resources and agrochemicals can induce negative impacts on the environment. The nutrients, water, light, and temperature can be classified as determinants in crop production. Most of the fertilizers are costly. Prolonged use of nitrogen can enhance methane and nitrous oxide emissions. Decreased nutrient uptake can result in the pollution of the coastal water and inland waters. This can be due to runoff and leaching. Most of the agricultural production systems need nutrients in excess and this is the main problem whilst studying the environmental concerns. The plant breeding methods can enhance crop yield. Yield can be considered as an important determinant in deciding the NUE or nutrient use efficiency (Barraclough et al., 1981).

The agricultural systems demand more inputs and they can be derived from organic products and inorganic products. The nutrient use efficiency can be based on two traits. They are NUpE or nitrogen uptake efficiency and NUtE or nitrogen utilization efficiency. The nutrient capture can be considered as a root trait and it needs sinks so that there will be temporary storage. These stored nutrients can be transported to other organs of the plant system. The acquisition of the nutrients will be dependent on the architecture of the roots and the functioning efficiency of the exudates and transporters. The plant response to the availability of the nutrients can be classified as complicated. There will be several pathway fluxes that need enzymes. The regulatory loops in the plant systems will be important when explaining the nutrient use efficiency.

The crop yield can be estimated by the production of the carbohydrates and their storage. When the crop yield is increased, most of the nutritional compounds will be diluted. The absorption of the mineral nutrients can be via transmembrane. The storage in the vacuole can be important (Kataoka et al., 2004). The efficiency of nutrient utilization can be determined or expressed in terms of biomass production.

5.2 ROOT SYSTEMS

The root systems can be used by the plants to extract water and nutrients from the soil. The depth of the roots can be about 2m. It was observed that the roots will follow the logarithmic pattern in the soils. At the surface the root concentration is high. *Arabidopsis* species was well studied and the regulation of the genes in their growth was understood well. It is important to know that all the plant system is important so that the response towards the nutrients can be perfect. The recording of these systems in relation to nutrient availability is hard.

The roots can absorb nutrients and water and they will respond to several edaphic factors. The depth to which the root grows and their distribution can be based on the genetic and environmental factors that are prevalent in the soil profile. Environmental and cultural conditions plays a vital role in defining the depth of the roots. The downward progression of the roots can be based on the season and other factors. The growth of the root system can be affected by the chemical properties of the soils. The soils that accommodate the root system should have enough mechanical strength so that the plants can get anchored properly. The soils must have pores so that air and water can enter inside. These pores are essential to plant growth. The regions that have dense soils can reduce the growth of the root system.

Figure 5.1: Root system.

Source:http://facweb.furman.edu/~lthompson/bgy34/plantanatomy/plant_root.htm

These densely packed soils can stop the growth of the root system and so the oxygen supply to the roots will be limited. The roots can deform the soils and this will reduce the mechanical resistance. The difference in the density of the soils can be affected by the root cap properties. The roots will respond to salinity, aluminum, pH, and chemical environment. The spatial configuration of the roots can be called as root architecture. There will be the formation of the root apical meristems. There will be an initiation and extension of the lateral root branches. There are several signals that can affect the root system and architecture. It was reported that the genes can control the growth and production of the lateral roots. There are some important stages in the formation of the lateral roots.

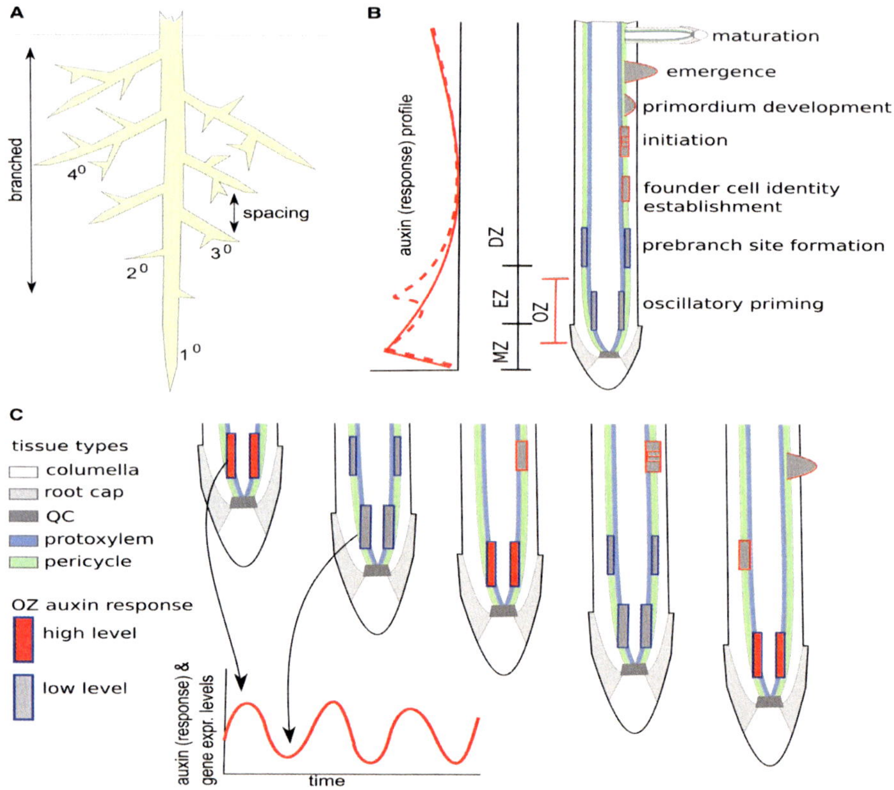

Figure 5.2: Lateral root meristem.

Source:http://www.plantcell.org/content/29/3/432

They are lateral root meristem (Figure 5.2) activation and elongation, the emergence of lateral root primordium or LRP, asymmetric cell division and stimulation of pericycle cells. The auxins can control the development of lateral roots. They can generate a signal that can regulate every stage of development. The elongation of the roots can be controlled by GAs or gibberellins. The architecture of the roots can be vital as it can determine the absorption of water and minerals. In the case of P deficient soils, there will be a decrease in the gravitropic sensitivity of the roots so that there will be shallow root growth. The root systems that are shallow in nature can absorb more P than the deep root systems. The root systems can have complex architecture and so the roots will be in contact with the nutrients that may be present in the soils. In order to uptake the nutrients, the plant should push themselves to the region where the nutrients are present or the nutrients must be dissolved in the soils. The root systems follow computational models. There are several research studies that deal with the transport of the nutrients that involves diffusion and mass flow. The mass flow can be due to convection and this happens because of transpiration. Most of the ions that are dissolved will be transported to the surface of the roots.

Nutrients are randomly distributed in the soils and it is always important to know the response of root towards nutrients. In presence of high nutrient concentration, the number of roots will be increased. The first order and the second order roots will be increased with the nitrate, phosphate, and ammonium ions. Two genes that can regulate the signal transduction pathway are ANR1 and AXR4. ANR1 gene can be classified as a nitrate regulator. AXR4 gene can be sensitive to the auxin concentration. The root tip has a network that can affect the growth rate of the roots. The nitrate molecules can stimulate and enhance the production of the lateral roots. If glutamate is present, then the growth of the roots will be hindered. There are several effects presented by the nutrients that can be associated with the growth of the roots (Brun et al., 2010). There are many works that are linked with the assessment of the root growth kinetics. The microbes in the rhizosphere and the roots can alter the structure of the soils. It was reported that 80% of the plant species can develop mycorrhizal associations and can exhibit symbioses (Smith et al., 2008). The hyphae of the mycorrhizae can be more effective. AM fungi or Arbuscular Mycorrhizae can help the plants that grow in the soils deficient with phosphates. There will be the transport of nutrients from the fungal species to the host plants. The mycorrhizae fungi can behave in different fashions and no two mycorrhizal associations are the same. The carbon compounds can get deposited in the rhizosphere area and this can be called as rhizodeposition. The energy stored in the rhizosphere region can be used by several microbes. These rhizodeposits can activate the growth of the microbes and they can help the plants in several ways. The uptake of phosphorus can be based on several root factors. They are solubilization effects, mycorrhizal effects, and root geometry. The variations in the solubility of P in the vicinity of the root surface can affect the P uptake. The variation in the infection rates and extent of infection can also affect the uptake of P molecules. The difference in the root length, root diameter, density, length of root hair can affect the uptake of P molecules. The root architecture and the soil conditions can affect nutrient uptake. There is a significant relationship between the P uptake and root architecture along with the P use efficiency (Figure 5.3). There is a genetic variation that determines the architecture of the roots (Hargreaves et al., 2009).

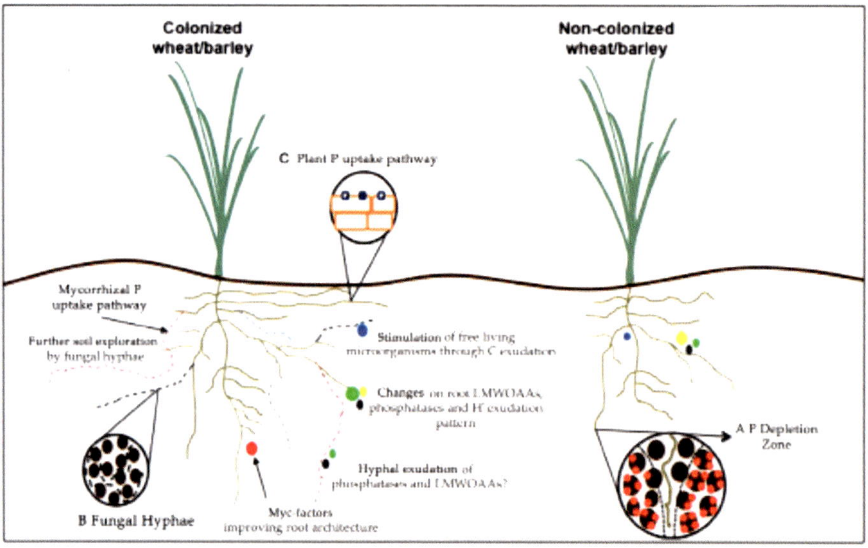

Figure 5.3: P uptake and roots.

Source:https://www.frontiersin.org/articles/10.3389/fpls.2018.00752/full

Several genetic factors that can control the structure of the root system were identified with the latest DNA based markers. Several traits will define the efficiency and absorption of the nutrients. It is important to detect the markers and traits that can promote the breeding mechanisms of the genotypes (Atkinson et al., 2015). The roots must contact with the fertilizer and these contact points can be important. The banded P can be utilized by the plant species and it has many uses. The root modification to enhance the absorption of the nutrients was reported. The phosphatase enzymes can be produced by the fungal and bacterial members that can cleave the organic P into inorganic P so that it will be available for the plants. The inorganic P molecules can be absorbed by the microbes. Some plants can grow in the P deficient soils that can produce phosphatase enzymes. The secretion of the phosphatases can be observed in the roots and it can be linked with the absorption of P. Wheat can use P from the diester forms and monoester forms. The genetic modification process can be applied to increase the production of the phosphatases. This can yield lesser benefits as this enzyme will be absorbed by the solids in the soils.

5.3 RHIZOSPHERE

The rhizosphere zone (Figure 5.4) can be important in determining the NUE or nutrient use efficiency in the crops. The rhizosphere can act as an interface between the soils and roots. The uptake of the nutrients can be based on rhizosphere conditions. The plants can change the availability of the nutrients in the rhizosphere zone. It is possible only with the production of the compounds that can mobilize the nutrients. The change in the pH values can also determine the availability of the nutrients. The microbes in the rhizosphere zone can enhance the availability of nutrients. This can occur due to activation of the root growth and nutrient mobilization. The plant roots can alter the biological and physicochemical conditions of the soils. The rhizosphere zone can be present just a few millimeters underneath. The variations in the roots can be affected by several factors. They are the age of the roots, type of soils, temperature, rainfall, nutritional status, and plant genotype (Neumann et al., 2019). The nutrient availability in the rhizosphere zone can be regulated by the cumulative effect of the microbial action, plant characteristics and soil properties. The structure of the soils can be stable in the rhizosphere as the roots secrete the exudates that are of high molecular weight. These exudates can work as glue that can make the soil particles bind together. The hyphae and root hairs can pack the soil particles. The high molecular weight exudates can be available to the microbes and so these chemicals can be decomposed easily.

The plant roots can absorb the inorganic nutrients that are present in the soil solution as dissolved ions. There will be limited availability of the nutrients that are close to the plant roots. A very limited portion of the major nutrients can be absorbed by the plants due to interception (Deubel et al, 2005). The movement of the nutrients in the soil can be important for the plants. The movement of the nutrients can be via diffusion and mass flow.

The concentration of the nutrients in the rhizosphere can be regulated by the uptake of the nutrients and movement of the nutrients. In case of excess uptake of water, then there will be a high rate of transpiration. This will lead to a reduction of water volumes in the rhizosphere zone. This happens mostly in the arid and semi-arid zones. The limitation of water can negatively impact the uptake of nutrients. The microbes present in the rhizosphere and roots can produce polysaccharides and these chemicals can bind with water molecules. Due to the adhesion of the water molecules, there will be a reduced rate of drying. Some of the deep-rooted plants can survive in rough climates. At night times, water can get absorbed by the roots. The hydraulic lift can be considered as an important adaptation measure to enhance nutrient use efficiency.

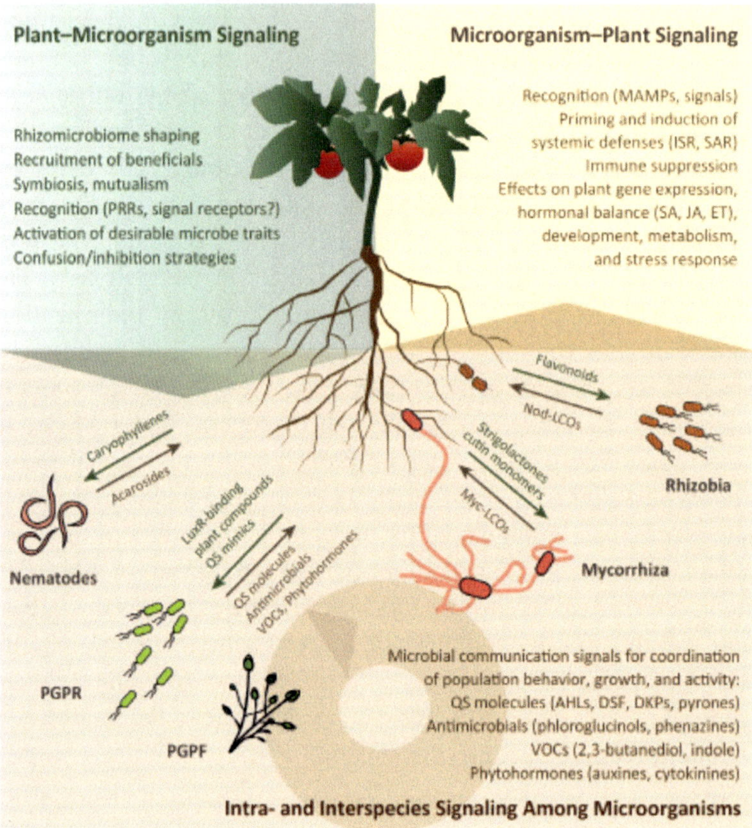

Figure 5.4: Rhizosphere zone activity.

Source:https://www.cell.com/trends/plant-science/fulltext/S1360-1385(16)00006-6

The microbes in the roots and rhizosphere zone will alter the redox potential. In the case of wet soils and submerged soils, there will be less oxygen and this will lead to declined redox potential. The availability of manganese and iron can be high in reducing conditions. These conditions will result in toxic conditions. Due to the presence of aerenchyma, the oxygen molecules can get transported from the shoots and reach the rhizosphere zone. Once the redox potential increases, there will be rapid oxidation of manganese and iron at the vicinity of the root surface. Any alteration in the pH value of the rhizosphere can affect the availability of the nutrients. In the case of acidic pH, there will be increased absorption of metals, iron, and phosphorus. It is important to know that the acidification of the rhizosphere can be classified as an adaptive response that was observed in many plants. This was evident in the case of phosphorus and iron deficiency. Mostly, the acidification of the rhizosphere will be due to the ammonium supply. It was observed that the nitrogen mineralization was increased.

Most of the aerobic soils will have less amounts of manganese and iron molecules. The microbes and the plants can enhance their absorption capacity of the iron and manganese ions by modifying the reductase activity of the plasmalemma. The Fe^{3+} can be reduced to Fe^{2+} and Mn^{4+} can be reduced to Mn^{2+}. These reductions can enhance their solubilities and due to this, their absorption by the microbes and plants will be increased. The acidification of the rhizosphere can also enhance the absorption rates. The chelating ions can also enhance absorption capacities. Some of the chelating compounds are phenolics, anions of organic acids and siderophores. The pH of the soils can determine the availability of the nutrients and so the nutrient efficiency of the plant. The roots can alter the pH values in the rhizosphere zones and carbon dioxide will be released. Due to this, there will be either uptake or release of the protons. The respiration process in the microbes and roots can lower the pH values of the rhizosphere. The carbon dioxide that was absorbed can produce H_2CO_3. The respiration process can lower the rhizosphere pH.

There are major changes in the pH of the rhizosphere due to the balance between the cation and anion uptake in the roots. Most of the cells should maintain balance in the charge so that the extra cations can be regulated. Alkalization can regulate extra anions. The acidification of the rhizosphere can enhance iron and phosphorus availability. The acidification of the rhizosphere can be classified as an adaptive response. An increase in the rhizosphere pH can lower aluminum toxicity. The alterations in the pH values can change the biofilm surfaces on the roots. The variations observed in the pH of the soils is due to the nutrient transformations. There will be enhanced nitrogen mineralization in the soils. The nitrification process can lower the pH values and ammonification can enhance the pH values. The nitrogen fixation that is observed in the legume plants can enhance the N_2 uptake and other crops can fix atmospheric nitrogen via crop rotation. The exudates released can be considered as an energy source for microbes that can fix nitrogen molecules.

Azospirillum species can be classified as a nitrogen fixer that is free-living and this species is observed in the soils. Bacteria can invade the root system and occupy the vascular systems of the stems and roots. The transformation of organic N into nitrate and ammonium can be called as nitrogen mineralization and this process occurs due to microbes. It was reported that 90% of the nitrogen present in the soils is organic. N mineralization can be high in the rhizosphere zone than in the soils that are tightly packed. There will be anaerobic microsites in the rhizosphere zones due to the denitrification process. Biofilms on the surface of the roots can be formed due to ammonia-oxidizing bacteria.

The microbes can lower the availability of N to the plants and this is possible with the microbial biomass. Most of the microbes exhibit low carbon to nitrogen ratios and so it was clear that these microbes need more nitrogen. The immobilization of the nitrogen molecules can lower the N availability. Most of the exudates secreted by the roots are having less N concentrations (Neumann, 2007). Most of the phosphorus molecules are not available to the plants. The phosphorus molecules have lower mobility. The P that is available can be reduced in the roots. The transporters in the roots can absorb the P molecules. The plants can be able to enhance the P uptake from the soils via altering the physiology and morphology of the roots. Increased length of the roots can be considered as an important morphological adaptation that can enhance the plant's capacity to approach excess soil volume. The growth of the root hairs can also affect the absorption process. The clover plants can have long root hairs that can help in the absorption of the P molecules. The hyphae of VAM fungi can also help the plant roots to penetrate deeper.

The physiological variations like secretion of phosphatase enzymes, phenolics, organic acid anions and protons can be important. The release of proteins can enhance the acidification process. This acidification process can enhance the solubility of the calcium salts in the basic soils. The iron and aluminum phosphates can get dissolved in acidic soils. The anions of the organic acids can get enhanced with the increase in P availability. There will be solubilization of the aluminum and iron molecules along with anion exchange. In case of high pH in the cytoplasm, the anions of the organic acids will be released and so the pH may not be decreased as expected. The phenolic compounds can be secreted by the roots with the enhanced availability of P. Most of the P can occur in the organic form as there is more organic matter.The performance of the exudates can be based on the pH levels of the soil. It is also based on phosphatases and anions (George et al., 2007). The density of the cluster roots can enhance as the availability of P declines. The phenolics can get exudates from these cluster roots. The phenolics can decrease the activity of the microbes (Weisskopf et al., 2006). The availability of P in the soils can be lowered due to P immobilization. Some of the compounds that can mobilize P molecules can undergo decomposition. The role of rhizosphere microbes can vary with the plant species. The iron content in the soils can be about 4%. The chelates can bind with iron molecules, zinc, and copper molecules. The chelating form of the iron molecules can be called as iron-phytosiderophore. These chelating forms can be produced at the root tips (Fan et al., 2011). The chelated ferric iron molecules can get reduced at the outside envelope of the cell and in some cases, the reduction can occur inside the cell envelope. Plants rarely use bacterial siderophores. The bacterial siderophores can exhibit a high affinity to iron molecules. Microbes can modulate the iron absorption process by the plants. The microbes in the rhizosphere region can lower the uptake of iron molecules. The microbes can use the chelators. Some of the plant-derived chelators can be decomposed by the microbial action. The microbes can produce chelators of high affinity. The uptake of the iron molecules can be based on the relative concentrations.

Most of the arable lands in the world are deficient in manganese. Mostly, the oxidation process occurs due to biological intervention and reduction occurs due to chemical or biological intervention. The crop genotypes can vary in their sensitivity against Mn deficiency. The increase in the Mn fertilizer can lower the infection rates in wheat crops. The roots that are Mn deficient exhibit minimal lignin concentrations. Once the microbial density increases, there will be a decrease in the secretion of the exudates and diminished uptake of nutrients. The nutrient use efficiency can be affected by microbial density and exudation. The plants can lower the activity of the microbes and so the competition among the microbes will be reduced. The biomass of the microbes in the roots can be increased due to infestation by protozoa and nematodes. The signaling compounds can attract microbes like nitrogen fixers and mycorrhizal fungi. These signaling compounds can also repel the deleterious microbes and pathogen. Sugars will be released from the plants to enhance the microbial solubilization and mineralization. Phytosiderophores can help the plants to enhance iron absorption. The anions of the organic acids can enhance the availability of metals and phosphorus. The change in pH can enhance the availability of the metals and P. The polysaccharides will be released by the plants to enhance the structural stability of the soils. This can be achieved by enhanced water transport and aeration. There will be a hydraulic lift in the plants that can enhance the movement of the nutrients. The enhanced movement of the nutrients can be due to mass flow and diffusion. The hydraulic lift can enhance microbial solubilization and mineralization. The growth of root hairs can enhance the depletion zone of the roots. Enhanced nutrient availability can result in mineralization and there will be a release of polysaccharides. Due to this, there will be a secretion of organic acid anions. There will be secretion of the siderophores. In case of diminished nutrient availability, there will be secretion of the growth stimulators like mycorrhizal fungi and roots. There will be pH alteration and release of nutrients from the biomass of the microbes. In case of reduced nutrient availability, there will be decomposed root exudates. Ligand exchange will occur in decreased nutrient availability. The nutrients will be mobilized because of the exudates released from the roots. There will be reduced colonization of the mycorrhizae and root growth.

5.4 CANOPY PHYSIOLOGY

The fertilizer made up of mineral nutrients can demand more production costs and so they can put a burden on the farmers. Due to leaching and other emissions, there will be several environmental impacts. Minimizing fertilizer inputs can be important in managing the cultivars. The canopy processes can increase nutrient use efficiency. The canopy traits can help in the operation of the nutrients. Some factors can affect the photosynthetic capacity and radiation interception. The photosynthetic capacity must be increased per unit of nitrogen. We need to optimize the nitrogen distribution vertically. Enescent and remobilization should be optimized. The crops that have enhanced nutrient use efficiency can provide monetary benefits to the farmers. This will lower the contamination levels of the environment. This will also reduce fertilizer inputs. The crops that exhibit minimal nutrient use efficiency can cause leaching so that the groundwater will be polluted. There will be eutrophication of the lakes and rivers. Excess application of nitrogen fertilizers can add to global warming. This is since N_2O emissions are released into the atmosphere (Goulding et al., 2009).

Breeding varieties were developed to enhance nutrient use efficiency. About 60% of the nitrogen is present in the chloroplasts. This was evident in the wheat crops. About 75% of the nitrogen that was reduced can be found in the cereal leaves. We can observe the nitrogen molecules in the mesophyll cells. The canopy can respond to the nitrogen stress and we can associate it with the grain yields. When the crops are deficient in N, there will be reduced photosynthesis per leaf, low stomatal conductance, reduced chlorophyll, and leaf N and lowered leaf area. Due to these factors, there will be lowered radiation interception. The radiation use efficiency will be affected (Foulkes et al., 2008). The overall photosynthesis mechanism will be hindered when the N is in low concentrations. There will be an abortion of the panicles and grains in maize due to reduced N concentrations (Monneveux et al., 2005). The photosynthetic capacity of the canopy should be increased. We need to optimize the nitrogen remobilization process. The senescence process should be properly addressed and abated properly. There should be an optimization of the vertical N distribution all over the canopy. We need to enhance the photosynthetic rate of every leaf if possible. The development of genetic engineering was linked with the increase in the harvest index, especially in wheat crops.

5.5 REMOBILIZATION OF THE NUTRIENTS

Some of the crops can be monocarpic and annual in nature. The living tissues can be important in the whole process of germination as in annual crops, only embryo and other tissues will be remaining. The research initiatives in the nutrient use efficiency and plant productivity can be get involved in the senescence mechanism. In the senescence process, there will be yellowing of leaves. After the yellowing process, there will be the formation of the necrotic tissues. Most of the senescence process can get initiated from the lower leaves and then reach the leaves that are present at the top. The yellowing of the leaf tissue can be due to the degradation of chlorophyll molecules. Due to the yellowing of the leaves, the chloroplastic capacity will be lowered so that the entire photosynthesis process will be disturbed. The development of the plants can be affected by the surrounding environment. The wheat crops will exhibit senescence due to hot and dry weather conditions. The temperature and water conditions along with the nutrients in the soils can be considered as important factors in determining the senescence. The reduction in the N concentrations can induce the plants to go for premature senescence. Logging problems can cause plants to go for senescence (Yang et al., 2005). The nutrient remobilization process can be considered as a core for the senescence process and this is directly linked with the nutrient use efficiency. Most of the plants that survive on minimal nutrients can be prone to pathogen attacks. The remobilization process can occur due to carbohydrate stores in the plants. The nitrogen content may not be transported to the seeds.

In the process of protein degradation and production, some of the ammonia molecules can reach the leaf tissues and this will encourage the senescence mechanism in the leaves. The ammonia molecules can be transformed into amino acids and this occurs via GS or glutamine synthetases. This process can be considered important in the N remobilization process (Tabuchi et al., 2007). In the maize crops, the role of GS in the remobilization of the nitrogen molecules was reported. The same process was reported in rice plants (Tabuchi et al., 2007). The cellular components that play an important role in the remobilization process are cell walls, membranes, and chloroplasts. There will be protein degradation in these cellular components through the ubiquitin pathway, autophagy, and proteases. There will be degradation of the fatty acids through beta-oxidation and lipases. There will be a production of amino acids due to protein degradation and Acetyl-CoA will be formed due to the degradation of the fatty acids. GS1 plays a major role in the production of amino acids and this can be noticed in the late senescence phase. The dominating amino acid in the senescence process is the glutamine. Most of the transportation processes can be based on tonoplasts. There will be activation of several genes so that the chelating factors can be activated. These genes are important in the remobilization process, especially in the senescence phase.

The death of the cells can abort the senescence phenomena and this process can be compared with the apoptosis process. The death of cells in the plants due to the senescence process can be classified as programmed cell death. In this process, the entire plant regulation mechanisms can get involved and it is also dependent on several hormones. The regulation of senescence can be based on hormones and other associated factors. This can be observed in the plants that respond to the drought stress. ABA plays an important role in this process. The production of ABA can be considered as an important signal if plants are subjected to water stress (Harding et al., 2009). The ABA molecules can be transported from the roots to the leaves via the xylem. At first, there will be the closure of stomata.

The transport and effect of ABA can be in acropetal order. The putative receptors and their related transport mechanisms are still debatable to this day (Pennisi, 2009). ABA can be considered as a link between the external environmental factors and stress with the reproductive organs and the vegetative portions of the plants. In some cases, the ABA molecules can be transported to seeds from the leaves. This is the reason why the concentration of ABA will be higher in the seeds (Nambara et al., 2003). It has a major role in the regulation of seed proteins. There is a strong link between the senescence process and the development of the seeds. The ABA molecules can get involved in several physiological processes. These compounds can also influence the production of ethylene and cytokinins. The ABA hormone can be classified as a sensory molecule in plants. The cytokinins can be classified as sensory molecules to know the estimate of the N status of the crops (Sakakibara et al., 2006). Ethylene can be classified as senescence-enhancing hormone. The addition or spraying of ethylene can activate the gene expression patterns. Antioxidants can exhibit a major role in the senescence process.

5.6 GRAIN COMPOSITION

The seeds can play an important role in the propagation of the plants. The seeds can be used in the sexual reproduction process. The germination of the seeds may demand more energy so that the building blocks can be produced. The building blocks will be produced until the seeds start growing independently. There are three groups of storage compounds. They are proteins, starch, and lipids. These compounds can be stored in the organelles. Some of the tissues that are used for storage are protein bodies, amyloplasts, and oil bodies. MMineral nutrition can affect the storage capacity. The minerals that are absorbed can determine the structure of the storage tissues and compounds. N and S can play an important role in the storage process. There is a relationship between the grain N and available N. The quality of the proteins can be defined by the amino acids. These amino acids may not be produced by humans.

Sulfur can be classified as an important component of the proteins that are produced from the amino acids. The amino acids that have sulfur groups are methionine and cysteine. The crops that grow in the United Kingdom can get sulfur molecules in two forms. They can get S in the form of ammonium sulfate or superphosphate. Some of the sulfur molecules can reach the crops through the atmosphere. The sulfur and nitrogen compounds can be supplied to the crop plants in the form of foliar sprays. The polymerization of the proteins can be controlled by the gluten molecules.

There should be a balance between the sulfur and nitrogen molecule. This balance can affect the levels of S-methyl methionine molecules in malt. This balance can also affect the sulfur content. Beta-glucan can enhance the quality of oats and make them more attractive to consumers seeking healthier food. It is always important to add nitrogen fertilizers so that the protein content in the grains will be enhanced. We can add foliar urea so that the risk of lodging will be reduced. When the grains of are barley are hydrolyzedthe aspartic acid is produced. Asparagine can enhance the protein content and there will be nitrogen fertilization (Claus et al., 2006). The amino acid asparagine can be present in the embryo and bran of rye plants and wheat crops.

Selenium can be considered an important micronutrient as it will help in the antioxidation process (Arthur et al., 2003). The sulfates and selenates can compete for the same transporters and enzymes. When the sulfur molecules are used to enhance the quality of bread, there will be a decrease in the selenium content in the grains (Stroud et al., 2009). The gene expression can be controlled by the sulfur and nitrogen levels of the grains.

5.7 GENETIC IMPROVEMENT

In the last century, breeding was never an option in the case of wheat crops to enhance the NUE. It was reported that the nitrogen fertilizers were cheaper then. The height of the plants can be reduced so that there will resistance against lodging and the crops can sustain the N fertilizers effectively. Most of the environmental impacts due to N fertilizers were minimized. It is important to select and produce high-yielding varieties so that there will be enhanced nitrogen efficiency. The N concentration will be high in the external environment or the atmosphere and the nitrogen content can be maintained so that the crop cycles must be maintained properly. The seasonal availability in France didn't match the demand for the crops. The N molecules can be considered as an important chemical in the grain proteins. The nitrogen concentrations can define wheat production and yields. The distilling and biofuel may need lesser amounts of proteins in the grain and so there will be high alcohol yields. The dry matter and the nitrogen content in the grains may not be linked with the nitrogen use efficiency. The wheat genotypes can absorb more N from the fertilizers and soils. The nitrogen use efficiency can be the ratio between the growth yields and nitrogen molecules that can be absorbed by the plants. The improvements in genetic engineering have enhanced the agricultural outputs. The abiotic and biotic environments can get altered as time progresses (Fischer et al., 2010). The nitrogen content that is present in the atmosphere is equally available to all the plant species.

The storage proteins can be classified into glutenin and gliadins. The glutenins can be classified as polymeric proteins. They have viscoelastic properties. The gliadins have monomeric properties. Transgenics can increase the molecular weight of the glutenins. Due to transgenic crops, we can even reduce the protein content. F_1 hybrids of wheat are being used in several countries. NUE can be classified as a complex trait and it is based on many genes. In the process of selection, it is important to detect the key regions of the chromosome and so the markers. Quantitative genetics was developed to detect the utilization and uptake of nitrogen in barley plants. With the latest advancements in the field of genetic engineering, the identification of chromosomal regions that code for nutrient uptake and recycling, vegetative growth and root growth became easier than before.

5.8 MOLECULAR GENETICS

The primary target of any crop improvement program would be yield and disease-free products. The increase in the advancements in the field of genetic engineering has led to enhanced yields (Miflin, 2000). Nitrogen was classified as a limiting factor in the production of crops. The overuse of the N fertilizers caused negative environmental impacts and this practice has forced the biosphere to absorb more and more carbon compounds. Excess utilization of synthetic fertilizers can reduce the organic matter in the soils and so the microbial populations will be declined. Most of the N fertilizers can emit nitrogen oxides when mixed with the air mixture in the croplands. The nitrogen oxides can combine with ozone gas and can result in the production of ammonia-rich acid rains that can be detrimental to the biosphere zones (Galloway et al., 2013). The improvement of the NUE can be considered as a challenge for decades. This can be due to the ever-growing demand for crop yields. Water pollution can result if there is excess N fertilization. It is often debated that there should be the production of hybrid varieties that can consume less N content. Research is in progress to link plant nutrition with plant development and vice versa.

The fertilizers that are being produced require more fossil fuels and due to this, there will be increased fertilizer costs. The grains from the cereal plants can provide nutrition to half of the world population. The novel crop varieties must utilize minimal N fertilizer and they must give good quality of grains and so the yield. The fundamental metabolic systems must be kept in mind while attempting the new genetic breeding techniques.

As the population rate increases, there will be a demand for agricultural products and so we need to safeguard the environmental parameters. The farmers are having economic pressure due to enhanced costs of the fertilizer. Due to the enhanced cost of fossil fuels, stress has been high on the biofuels (Walker, 2010). There are many tools that can improve the plant's NUE..It is important to enhance the crop quality at the limited expense of the fertilizers. This approach can reduce the negative impacts on the environment. The vegetative growth phase and remobilization phase are important in managing the nitrogen metabolism. In the younger stages, the roots and leaves can act as sinks or sink organs. These sink organs can store N in its inorganic state. The amino acids and proteins can be stored in these sink organs. In the case of senescence, all these inorganic N contents will reach the portions where the growth initiates.

There are root-specific transporters that can permit the uptake of the ammonium molecules. The assimilation of the inorganic nitrogen can be an important limiting factor that can affect productivity and biomass. Most of the processes that entangle with the NUE can be based on individual factors or a collection of individual factors (Masclaux et al., 2008). Most of the processes can occur in parallel with each other. It is always important to study transcriptional regulation and post-transcriptional changes. These changes can be linked with nitrogen reduction and associated processes. The impact of molecular modification and genetic modification must be studied (O'Brien et al., 1985). The important goal in the production of transgenic plants is that there should get enhanced NUE and that is possible only through overexpression of enzymes and proteins. Through the latest genetic engineering technologies, a minimal increase in protein was observed in rice crops (Zhou et al., 2009).

In the case of pine plants, there are two GS genes that can encode GS1a and GS1b. The Dof1 TF or transcription factor may not be important in the regulation of carbon and nitrogen metabolism. It was observed that enhanced NADH-GOGAT can increase the weight of the spikelet. Quantitative trait loci (QTLs) are considered as an important tool in the identification of structural and regulatory genes that can get involved in the agronomic and physiological processes. Positional cloning can be considered an important technique that can target the chromosomal regions (Salvi et al., 2007).

The QTL approach can help us to understand the interactions between the regulatory elements). QTLs can be studied in relation to environmental changes. The plant productivity can be based on the genetic variability. Understanding the genetic variability of the nitrogen non-limiting and limiting conditions can be considered as important as it is directly associated with NUE.

Chapter 6

Metabolic Networks

CONTENTS

6.1 Introduction 118
6.2 Photorespiration 119
6.3 Measurement of Metabolites 122
6.4 Enzymes 127
6.5 Isotopes 128
6.6 Topology 128
6.7 Dynamic Flux Analysis 130
6.8 Primary Metabolites 130

6.1 INTRODUCTION

The metabolism can be self-regulating, highly organized and is an uninterrupted process that comprises small molecules, chemical modifications, uptake, and transport of nutrients. The metabolism process can be explained in two primary quantities. They are flux and concentration. Flow can be classified as potential and, in some cases, the reverse is also possible. The enzymes and metabolites can interact with each other via several interactions.

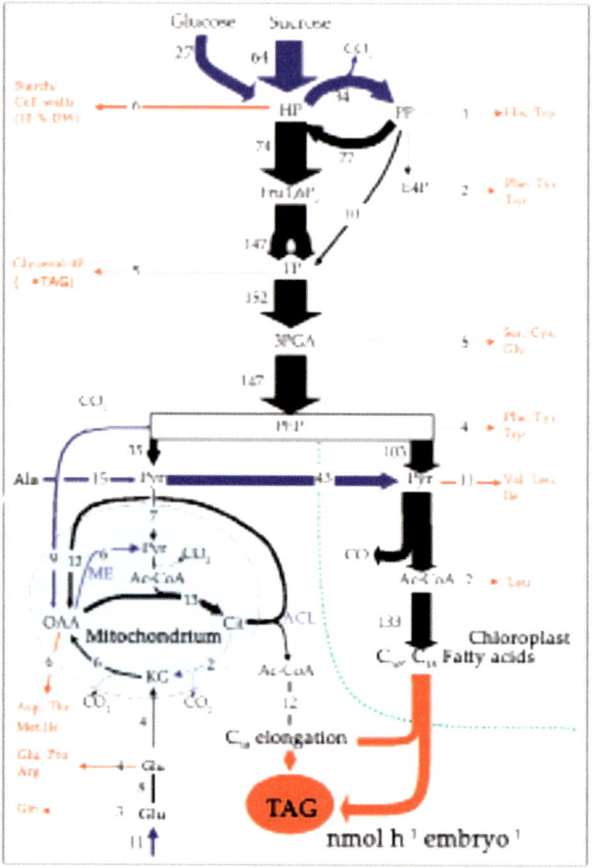

Figure 6.1: Plant metabolic networks.

Source:https://www.semanticscholar.org/paper/Understanding-flux-in-plant-metabolic-networks.-Schwender-Ohlrogge/bda1c7f5bb71abd030e625b-3fe146a6587db02fe/figure/0

The prediction models can be considered as the latest advancements to determine the metabolic pathways at the molecular level (Sweetlove et al., 2009). Most of the properties of the system must be studied as a whole structure and not as isolated entities. It was established that there are many levels of the organization and some properties emerge as the complexity of the organization increases. The metabolic networks (Figure 6.1) can get surfaced in the case of centrality, metabolic flux and other elementary flux models evolved.

We can determine the metabolic networks in an accurate and precise way. It was considered that the basic property of the plant metabolic networks can be a selective partition of the metabolites that can inhabit organs, tissues, and cells. There are several transport mechanisms that can guide the metabolites in the correct direction. Most of the transporter molecules get involved in the metabolic processes that can be inside the cell and in some cases between the cells. The transporter molecules can be important in long-distant transport between the tissues. Most of the metabolic pathways can be partitioned between the organs, tissues, cells, and organelles. It is always important to consider long-distant transport processes and intracellular transport as crucial as most of the biochemical steps concerning growth, catabolism and biosynthesis depend on them.

6.2 PHOTORESPIRATION

The Rubisco or Ribulose Bisphosphate carboxylase oxygenase can be classified as a bifunctional enzyme and it is important in the carbon assimilation process of photosynthesis. This enzyme can get involved in the non-productive oxygenation and productive carboxylation process. These processes generally take place in presence of RuBP or ribulose 1,5-bisphosphate. Due to the oxygenation of RuBP, there will be the production of 3-PGA or 3-phosphoglycerate. There will be the production of 2-PG or 2-phosphoglycolate. This molecule is classified as toxic. One 2-PG molecule and one 3-PGA molecules can get involved in the photorespiration process. The toxicity of 2-PG can be eliminated and gets recycled as 3-PGA (Reumann et al., 2013). In this process, oxygen molecules will be used and CO_2 will be produced in presence of light. This can be called as photorespiration (Figure 6.2). The 2-PG molecules must be subjected to reaction pathways that are highly partitioned and compartmentalized. These reactions can produce 3-PGA molecules.

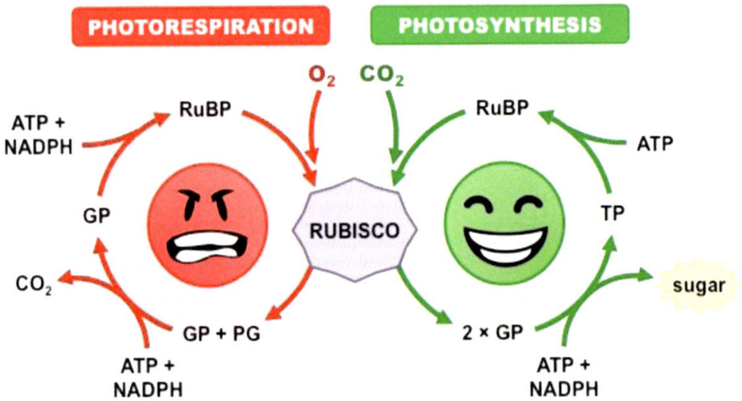

Figure 6.2: Photorespiration.

Source: https://ib.bioninja.com.au/higher-level/topic-8-metabolism-cell/untitled-2/c3-c4-and-cam-plants.html

In this pathway, there will be minimal loss of carbon dioxide that can be fixed. Due to this, there will be reduced deprivation of the intermediate molecules that generally gets produced in the Calvin cycle. This photorespiration pathway can be classified as a coordinated network. This network consists of about 14 enzymes that can be soluble. These enzymes can be present in between the chloroplasts, cytoplasm, mitochondria, and peroxisomes of the leaf. There will be 12 steps concerning membrane transport. The carbon cycle in the photorespiration process can start at the stroma of the chloroplasts. The phosphate glycolate phosphatase or PGP catalyzes the dephosphorylation of 2-PG. Glycolate formed can exit the chloroplasts via antiporter. These molecules will be absorbed by the peroxisomes. The glycolate molecules can be oxidized in the matrix of the peroxisomes. FMN-dependent glycolate oxidase can catalyze this oxidation process in the peroxisomes. Hydrogen peroxide will be produced as two electrons from the glycolate are shifted to oxygen molecules.

The glyoxylate can be modified via the transamination process. There are two aminotransferases that can catalyze the transamination reactions. The two aminotransferases are SGT or Serine:glyoxylate aminotransferase and GGT or Glutamate:glyoxylate aminotransferase. The glycine molecules can be absorbed by mitochondria. This process or transport can be mediated by the gly/ser transporter. In the matrix of mitochondria,two glycine molecules can be transformed into NADH, serine, NH_3 and CO_2 molecules

(Oliver et al., 1995). Glycine dehydrogenase can mediate this reaction. The serine hydroxypyruvate can also catalyze this reaction with glycine dehydrogenase. The serine molecules that are produced in mitochondria can be absorbed by the peroxisomes. At this stage, the serine molecules lose amino group via SGT. There will be the production of hydroxypyruvate. The hydroxypyruvate after consuming NADH can get converted into glycerate molecules. This reaction can be catalyzed by hydroxypyruvate reductase or HPR. There are two well defined metabolic transporters that participate in the photorespiration process. They are glutamate/malate translocators and plastidic 2-oxoglutarate/malate translocators (Weber et al., 2006).

Oxygenic photosynthesis and bio-geochemical cycles modulatethe atmosphere and reduced carbon dioxide levels with increased oxygen levels. Due to this, there was a selective pressure on the photosynthetic organisms. These organisms participated in the carbon concentrating mechanism or CCM. Due to this, there was reduced oxygenase activity and so the photorespiration also got lowered. Due to CCM, many species of algae and fungi were evolved (Thyssen et al., 2009). There are some plants that started adapting to C_4 photosynthesis. The CCM that is present in the C_4 plants reduces the limitation of carbon dioxide. The photosynthesis process in the C_4 plants can be classified as efficient. This process stands effective in the case of drought stress, salinity stress and hot temperatures.

Most of the C_4 plants can exhibit a high rate of nitrogen use efficiency. The specialized plant anatomy called Kranz anatomy plays an important role in the photosynthesis process of C_4 plants. The mesophyll cells contain chloroplasts and they are enclosed with bundle sheath cells or BSC forming a second vascular bundle. Kranz anatomy may not be mandatory in the photosynthesis process of C_4 plants (Voznesenskaya et al., 2001). Almost all CAM and C_4 plants have carbon dioxide in the form of bicarbonates. These molecules can get accumulated via PEPC or PEP carboxylase. The fixation of carbon dioxide can occur in the mesophyll cells (Cytosol). The oxalo acetic acid is absorbed by the chloroplasts that are present in the mesophyll cells. The Oxalo acetic acid or OAA can be subjected to reduction and there will be the formation of malate molecules. MDH or $NADP^+$-malate dehydrogenase enzyme mediates this reduction process. Then, the malate molecules that are formed will be sent back to the cytosol of the mesophyll cells. The malate molecules can reach bundle sheath cells. There will be a decarboxylation and oxidation process in the chloroplasts of BSC. Then, there will be the production of pyruvate molecules, carbon dioxide and NADPH. The carbon dioxide molecules that are produced in this step can be fixed again by the Rubisco enzyme. The pyruvate molecules will get phosphorylated. There will be the regeneration of an acceptor for carbon dioxide i.e. PEP. The enzyme that can catalyze this process can be called pyruvate phosphate dikinase or PPDK. The PEP molecules will reach the mesophyll cells and there they will act as acceptors for carbon dioxide. The malate molecules can be called bucket molecules for carbon dioxide. Some of the metabolites present in this bucket can interchange the carbon dioxide molecules from mesophyll and BSC. It can be assumed that the malate molecules are vehicles. The symplastic mode of diffusion can be based on the concentration gradients between the cell types. There will be different gradients of malate, triosephosphate and 3-PGA. The pyruvate gradient will be minimal. The pyruvate molecules can be sequestered in the cellular portions. The pyruvate concentration in the cytosol can be lowered. The transporters that can mediate the intermediates in the C_4 photosynthesis are still to be analyzed in detail. The intercellular transport can be aided by the diffusion process (Weiner et al., 2007). The BSC and mesophylls are linked with plasmodesmata. TPT or triose phosphate translocator can aid in the intracellular transport of 3-PGA and triose phosphate. The transport of these molecules will be between the plastids. In maize plants, these transporters were analyzed and documented. The TPT can mediate the counter-exchange process of 3-PGA and triose phosphate and in some cases inorganic phosphate. The transport of the pyruvate molecules was studied in the mesophyll cells of maize plants.

The metabolic channeling may not be restricted to the catalytic reactions that occur due to enzymes that are mostly soluble. This has become evident in the biosynthesis of arginine. The metabolic channeling was observed in the amino acid degradation (Islam et al., 2007) and biosynthesis of cysteine. In the case of yeast, enolase can be considered as a part of a macromolecular complex. This complex contains enzymes of the TCA cycle, mitochondrial family, and other metabolite transporters (Giege et al., 2003). The peroxisomes can accommodate malate, glutamate, serine and glycolate molecules. The transport of metabolites in cytosol and plastids can be important whilst studying the metabolic networks in plants (Weber et al., 2010).

6.3 MEASUREMENT OF METABOLITES

The metabolites can be classified as products that were formed as a result of enzymatic reactions that mostly takes place in cells. The metabolites will be produced in the cell so that the basic needs of the cell will be met. The metabolites can define the chemical reactions that occur in the cells. Due to the activity of many genes, there will be a production of the metabolites. The variation in the metabolite concentrations can be due to changes that occur in the gene expression. The quantitative analysis and comprehensive analysis of molecules in any biological system can be defined as metabolomics (Fiehn, 2001). There are many compounds that are being produced by the plants. It was reported that at least 200000 chemical compounds were present in the plant kingdom that exhibited variations in their reactivities and solubilities. About 50000 chemical compounds were characterized completely. These compounds were produced in different stages of plant's life cycle (Williams et al., 2007).

The tools needed to identify and measure the metabolites are mass spectroscopy, capillary electrophoresis, liquid chromatography, and gas chromatography. Some of the compounds can be measured without the aid of chromatography. Instead, we can use NMR or nuclear magnetic resonance spectroscopy and FT-ICR-MS or Fourier transform ion cyclotron resonance mass spectroscopy (Ratcliffe et al., 2010). There are many chemicals that have different properties. In this case, we can use a mass spectrometer to demarcate the properties of the samples. The separation of ions can be observed by mass spectrometry. In this technique, the compounds should be vaporized and then ionized. The main problem with mass spectrometry is that the samples will be subjected to ionization. We can use chemical ionization and electron impact ionization in the case of volatile compounds. There are three important techniques that can be used in the ionization process. They are matrix-assisted laser desorption/ionization or MALDI, atmospheric pressure chemical ionization or atmospheric pressure photon ionization (APCI/APPI) and electron spray ionization or ESI.

The plasma sources will be inductively coupled and we can use it to analyze the metals that are present in many samples. Some of the other techniques that can be used are thermal ionization, secondary ion mass spectroscopy or SIMS, desorption/ionization on silicon or DIOS, thermospray, fast atom bombardment or FAB and glow discharge. In this case, the ions formed will be represented by mass and charge. It can be assumed that the ions can exhibit different directions and speed in the magnetic and electric fields. The ions may get accelerated and can be separated in the magnetic and electric fields. All the separation process of the ions will be done in the mass analyzer. In order to achieve this, we can use different types of mass analyzers. In the biological applications, we can use FT-ICR-MS, quadrupole or Q and time-of-flight analyzer or TOF. We can use the TOF analyzer for this purpose as it is more convenient. In this analyzer, we need to use an electric field so that the ions can be accelerated. The time required by the ions to reach the detector can be measured and analyzed. It can be assumed that most of the particles have the same charge potential. This suggests that their kinetic energies are almost equal. The velocity of these particles will be based on their respective masses. With this technique, we can infer that the heavier ions take more time to reach the detector than the lighter ones.

We can use a quadrupole mass analyzer to stabilize and destabilize the ions. This can be achieved by supplying radio frequency or RF quadrupole field. The cloud of ions can be captured by magnetic and electric fields and this mostly occurs in the vacuum. Paul trap and Penning trap can be used to trap the ions. The most complex mass analyzer is the FT-ICR-MS. In this technique, we can know the mass and charge ratio of the ions produced. We can get this ratio by measuring the rotation frequency of the ions that exist in the magnetic field. In this case, there should be a homogenous magnetic field. The ions that were produced will be trapped in the magnetic field with the trapping plates. Due to the magnetic field, ions will have a circular motion. This motion will be perpendicular to the magnetic field. If we apply an RF pulse, there will be excitation of ions and so we can observe extended circular motion. This can be called as cyclotron frequency. The ions with different properties may not be identified at diverse places as some of the instruments may not be that sensitive. Each analyzer can come with flaws. In the case of a mass spectrometer, there will be two mass analyzers. This can be evident in the tandem mass spectrometry.

There are many steps in the tandem mass spectroscopy technique. The steps are linked with mass analysis and selection. The mass will be separated from the fragmentation process. We can use tandem mass spectroscopy in several rounds of MS. Each mass can detect a compound and its molecular weight.

The second mass analyzer is involved in the stabilization of the ions. This analyzer can collide with the gas. There will be an initiation of the fragmentation process. The third mass analyzer can record the fragments produced. We can use mass spectrometry in metabolomics. This spectrometry can be classified as a detector system. There will be a distinguished separation procedure. Most of the extracts of the metabolites can be derived from the biological sources. These metabolites can come with high rates of complexity. The complexities must be reduced so that the analyses will be simplified. The mass spectrometry may not support the compounds that are complex in nature. It is always important to know that the samples must be simplified prior to their injection into the mass spectrometer.

We can use liquid chromatography or gas chromatography for the separation of the complex molecules into simpler ones. The gas chromatography can be associated with mass spectrometry and this amalgamation was proved worthy with several chemical analyses. The GC-MS complex comes with several advantages and we can easily use this technique with limited expertise. It is economically cheaper. The GC-MS is being used since 1964 for several qualitative and quantitative analyses. The components will be resolved via differential partitioning between the stationary phase and mobile phase. The stationary phase will be solid in nature generally. The mobile phase will be the gaseous phase. Most of the samples that are to be used in GC-MS should be transformed or converted into liquid phase or solid phase and to the gaseous phase. This process can be called as the volatilization process. This process takes place at high temperatures. The samples will be subjected to 250°C. The samples will be allowed to pass via columns that have a stationary phase. The compounds should be allowed to undergo the ionization process. The GC-MS technique can be used to analyze compounds that can withstand high temperatures. In GC-MS analysis, the plant materials will be homogenized. Then, we need to extract the compounds with methyl alcohol. We need to maintain a minimum of 70°C. The insoluble compounds will be separated. The insoluble will be transferred into a dry aliquot. Then the compounds are subjected to derivatization. Post derivatization, we need to add samples to the GC-MS. We can analyze sterols, fatty acids, organic acids, and amino acids. In another pathway, the insoluble material can be cleaned up with chloroform.

The LC-MS can be adapted with many latest technologies like HPLC and others. These technologies can be used in the case of low-pressure systems. The separation of HPLC can be achieved by combining the diode-array detectors with LC. We can enhance the selectivity and specificity of the system with the aid of mass spectrometry. This will allow us to enhance the sensitivity in detecting the structural properties of the samples that can be present in the complex matrix. In the solute transformation process, we can use the atmospheric pressure ionization process and electrospray ionization techniques. LC-MS can be classified as a preferred technique to GC-MS. In case, the samples are highly concentrated, there will be matrix effects. There will be ion enhancement and ion suppression. The LC-MS technique has several advantages in both the separation and analysis of the samples that were extracted from the plant extracts (Sabu et al., 2002). It is always important to know that there are less mass spectral libraries that can be used in the identification of the peaks (Moco et al., 2006). The capillary electrophoresis can be combined with MS. It can be linked with LIF or Laser-induced fluorescence detection. This can be classified as a sensitive and accurate method. With this technique, we can understand the profiling of the extracts derived from different plant species. In this case, the use of a solvent is less. This is very less when compared with LC-MS. In this method, the derivatization process may not be needed. The runs can be faster relatively. There will be the production of ions already and so there is no need for repeated ionization is.

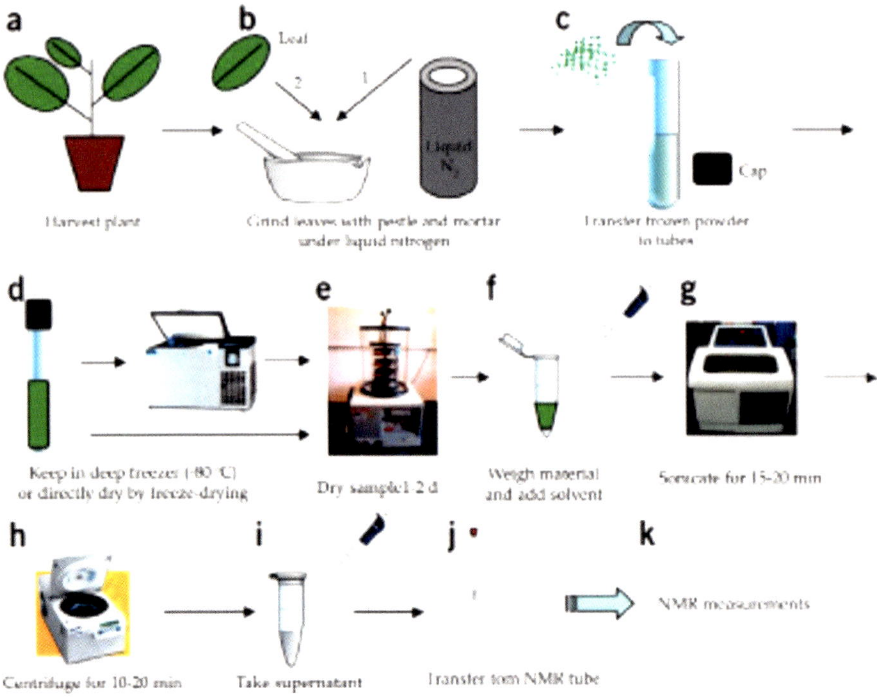

Figure 6.3: NMR technique.

Source:https://www.nature.com/articles/nprot.2009.237

NMR technique (Figure 6.3) can be used while studying the plant metabolomics. This technique can be classified as a non-destructive method. We can identify many metabolite classes based on stability, volatility, charge, and size (Dunn et al., 2005). With the NMR technique, we can answer several biological questions that can throw light on several processes in the plant systems. In this technique, we can use the magnetic properties that atoms exhibit. These properties are important in determining the structure of the chemical compounds. In this technique, radiofrequency pulses will be associated with strong magnetic fields. Due to this, there will be the production of spin states with higher energy in the nuclei. This can be observed in the atoms with odd atomic numbers and mass numbers like ^{13}C or ^{1}H. When the radiation is lost, the excited nuclei will return to its ground state or we can say low energy spin state. This radiation can be detected and based on this; we can construct the structure of the analyte. As NMR is a non-destructive analyzing tool, we can use it to study the metabolites in the living tissues. We can get structural information with high resolution. The NMR spectroscopy can be linked with the isotopic labeling technique and the resolution will be increased greatly. The flux of the compounds present in the plant extracts can be resolved easily. We can say that NMR has lesser resolution than other techniques of similar nature.

6.4 ENZYMES

The enzymes can enhance the rate of reaction. In the case of enzyme kinetics, we can study the measurements with mathematical treatment. The regulation of the enzymes can be due to products and substrates. The controlling of the enzymes with immediate product and substrate is not always possible in the case of metabolic pathways that are complicated. Most of the molecules can be generated via reactions that occur in the living systems. The regulation of the enzymes can be due to the allosteric effect. Most of the allosteric enzymes can combine with the substrates. The ligand can bind at a site that can cooperate in the formation of the enzyme-substrate complex. These allosteric enzymes can play a vital role in several metabolic pathways. Most of the Michaelis-Menten enzymes need increased substrate concentrations. The activity of the enzymes can be defined by calculating the concentration of product and substrate at given pH and temperatures. There are many principles that can permit the measurements of variations of products and substrates. Among several tools, luminometry, fluorimetry and UV-Visible spectrophotometry are important). We can use radioactivity methods to enhance resolution and sensitivity. The mass spectrometry techniques can be used in analyzing the enzyme activity (Greis, 2007). We can use electrochemical detection techniques to know the ion generating reactions. Biosensors can be used to enhance the resolution of the analyses. We can combine biosensors with microfluidics.

6.5 ISOTOPES

In plant research studies, radioactive isotopes were used for decades. We can use the isotope labeling techniques to detect the pathways. The labeling tools can be used to calculate the flux rates that occur in the case of reversible reactions (Wittmann et al., 2016). The labeling process can permit us to know the metabolic pathways in the plant system. The preparation of the samples can be classified as an important step in the entire process. This analysis needs tools that can be expensive. We can use 80% ethanol to extract the polar components with low molecular weight. Organic acids, amino acids and sugars can be subjected to fractionation with the aid of anion and cation columns (Krishnan et al., 2004).

6.6 TOPOLOGY

Most of the metabolic networks can be represented as graphs. The metabolite network can be any network that has substances only. These substances can be called as metabolites. The enzymatic reactions can be eliminated from the networks and any two metabolites can be linked. These metabolites can be linked with the same enzyme. The enzyme network can constitute enzymes only. These enzymes can catalyze several reactions. These metabolites can be eliminated from the enzyme network. In this case, two reactions can be linked. If the same substances get linked with the same enzyme, we can call it a metabolic network.

In the metabolic network, it is not always important that two metabolites need to be connected directly. We can establish these metabolite profiles with the aid of these metabolic networks. In this area, the clustering coefficient can be employed to determine the connectivity that may be present in the graphs. We can define this connectivity either locally or globally. The metabolic networks can be common in the technical and biological networks. The metabolic networks can be called as enzyme networks or metabolite networks. Most of the metabolic networks can be seen in the minute world networks.

The distribution of these metabolic networks was explained by Jeong and his co-workers (Jeong et al., 2000). There are a limited number of metabolites that can be linked with hundreds of metabolites. The network structure can be important in the evolution process. Most of the global structures of the metabolic networks can be determined by bow-tie structures. The carbon atomics concept was proposed by Arita. The candidate target genes and their prioritization can be considered whilst explaining the metabolic networks. The metabolites were ranged in the E. coli research studies. It was stated that the central metabolite can be pyruvate. The metabolites were ranked according to the minute metabolisms occurring enzymatically. The centralities like centroid values, closeness and eccentricity were considered similar. These centralities were applied to explain the metabolic networks of the *E. coli* bacterium. These centralities need to explain the role of the metabolites and the age of the metabolites. In the present scenario, ADP and inorganic phosphate are considered as having higher ranks. The drug target identification was considered as an important application whilst determining the metabolic networks. There are many software's that was used in determining the biological metabolic networks. Some of the popular ones are VisANT, Osprey and Cytoscape. The presence of complex networks in the technical and biological domains can be useful in network analysis. This analysis can help us to illustrate the structure and functional aspects of the metabolic networks. The local metabolic networks and global metabolic networks can be interleaved with the network motifs. These motifs are basically subgraphs that can help us understand the interconnections. These motifs can be considered as basic blocks of the networks that are much more complex. There are many motifs that needs to be identified. Some of the motif properties were studied in detail. Some of the motifs studied were the bi-fan motif and feed-forward motif (Nessler et al., 2007). The proteins present in the yeast have peculiar motifs that were conserved in the evolution process (Wuchty et al., 2005). There are some motifs that can get overlap. In this case, we can isolate the minor part only. We can classify the complex structures with motif clusters (Dobrin et al., 2004). They can be studied in the form of network themes and motif generalizations.

Metabolic networks can be classified into modules. These modules can coordinate with many diverse functions. In the case of cluster analysis, the objects will be grouped according to their properties. The cluster analysis can be employed in the bioinformatics sector. We can use this technique in pattern recognition, machine learning and data mining. This technique can be used in computational and statistical analysis. These techniques can be classified based on model-based methods, grid-based methods, density-based methods, hierarchical methods, and partitioning methods. The cluster analysis and multivariate analysis can be used in the analysis of metabolic networks (Jain et al., 1996). In the partitioning method, the objects will be divided into clusters. The clusters that were formed can be on-overlapping. We can use these partitions to simplify the criterion process. We can iterate this process until we get a perfect partition.

In the hierarchical method, the dataset that was obtained will be subjected to partitions but in a sequence or order. We can use dendrograms to show the hierarchical clustering tree. The network clustering method can comprise of clustering of data. This clustering can be shown in the networks. The network clusters can be represented as modules. There are many techniques that are involved in the network clustering process. This type of clustering can be seen in many biological systems (Spirin et al., 2003), food webs, metabolic networks, and signal transduction networks (Holme et al., 2003).

6.7 DYNAMIC FLUX ANALYSIS

The metabolic networks comprise of metabolic pathways. Most of the pathways are biochemical in nature. The metabolic flux analysis can be used in analyzing the microbial systems. The metabolic flux analysis can be associated with the isotopic labeling process. This technique can be useful in analyzing choline synthesis. The dynamic flux analysis can be used in measuring the labeling patterns of the metabolic patterns. We can use radiolabeling techniques as they have many uses in plant metabolite analysis. The analysis of structures can be used in analyzing the metabolites and detecting the conservation motifs.

6.8 PRIMARY METABOLITES

The primary metabolites in the plants can be classified as organic compounds and they are important in the reproduction, development, and growth of the plants. These metabolites can be considered as intermediates in several biosynthetic processes and in photosynthesis . The primary metabolites are lipids, steroids, fatty acids, nucleotides, amino acids, organic acids, and carbohydrates. These compounds can vary with plant species. These metabolites can be present in almost all plant species. These compounds are important in plant growth and development. There are about 10000 compounds that are produced in the plants (Pichersky et al., 2011). About 40% of the analytes can be called as metabolites. There are about 40 metabolites that were characterized in *Arabidopsis* species. In this process, NMR can be used in constructing plant profiles. We can compare these plant profiles and do a query. After performing the query process, we can visualize the data. We can perform statistical analysis and discriminate the profiles. After comparing with the standard profiles, we can identify the compounds.

The Metabolomic repository of Bordeaux is also called as MeRy-B. This repository stores metabolomic profiles of several plant species. 1H NMR was used to generate these metabolomic profiles. The repository comprises of experimental design, analytical metadata, spectral data, and compounds. The experimental design constitutes protocols, development stages, genotypes, experiments, projects, and biological sources. The analytical metadata comprise of protocols, extraction method, techniques, and instrumentation. The spectral data comprise peak lists, processed spectra data and pre-processed data. The compounds portion can constitute quantifications and other identified compounds. This platform can manage most of the data derived from NMR assisted metabolomics. There will be information about all the metabolites that were studied. We can easily determine the compounds that are present in the sample solutions. This MeRy-B stores about 40 metabolites along with NMR spectral data.

Polyphenolics can be considered as important metabolites that can be produced in the plants. These compounds will have antioxidant properties and most humans can consume these compounds due to their health benefits. Most of the metabolomic analysis is based on the mammalian samples and plant samples. It is based on liquid chromatography linked with electrospray ionization and MS. We can enhance the sensitivity and accuracy of this technique with the addition of isotopes. The isotopes can be linked with this technique before analysis. The isotope assisted samples can enhance the spectral analysis. This can greatly improve the estimation of the phenolic compounds in biological matrices that are often complex. Analysis of metabolites in the single cells can be possible via mass spectrometry. This technique can be critical as there will be less volumes that will be available. There will be more complex aspects. The analysis of the individual cells can be done by quantitative metabolomics. This analysis will help us to get information about the heterogeneity that can be present in the cells. The MS technique can be used in the quantitative analysis of the metabolites. With this technique, there will be enhanced selectivity and sensitivity. We can use the Laser ablation electrospray ionization technique or LAESI to analyze the metabolites that were derived from minute cell populations and biological samples. The phenotypic heterogeneity can be achieved in the isogenic cells. This can be done with quantitative metabolomics.

Mass spectrometry can be used to analyze many metabolites at a time. Matrix-assisted laser desorption was used to analyze cell lysates. ESI or Electrospray ionization was also used in the analysis of metabolites and this technique was combined with capillary electrophoresis. The quantitative analysis of the metabolites can be important in understanding the demarcations between individual cells. Mostly MS technique will be linked with liquid chromatography. In the quantification process, we need to perform multiple reaction monitoring or MRM. The quantification process can be done by analyzing the signal intensities. The small molecules can be analyzed perfectly with the aid of isotopes in the analysis. The desorption electrospray ionization or DESI can be used to analyze small molecules. The volatiles that are secreted by the plants are very important. They can get involved as hormones, defensive compounds and signaling molecules. The chemodiversity that was observed in the plants can help plants in tolerating both abiotic and biotic stresses. These chemicals can help the plants to adapt to rough environmental conditions.

The metabolites present in the plants can be classified into primary metabolites and secondary metabolites. The primary metabolites are fatty acids, sugars, and amino acids. The secondary metabolites can be used in reproduction, development, and other growth patterns. There are about 1700 compounds that were characterized (Knutson et al., 2012). The volatile organic compounds that are secreted by the plants can exhibit high vapor pressures and low boiling points. This is the reason why the volatility is higher in the plant species. The volatile organic compounds can help the plants in the pollination process. They help to attract parasitoids, predators, and pollinators (Yogendra et al., 2011). These compounds can inflict a defense mechanism in the plants. They can act as a platform for signaling. Chemical mimicry was also observed in the plants to attract predators. The volatile organic compounds can be classified as inducible and constitutive in nature.This can represent the ecological progress of the plants. These compounds were classified as indicators to represent stress in the plants, especially in greenhouses. These compounds helped in the integrated pest management as we can add suppressants and other chemicals to control the pests in the greenhouses (Jansen et al., 2011). A volatile collection trap is being used to quantify the volatiles in the plants.

The cellular metabolites can present spatiotemporal dynamics and it is hard to monitor all the parameters with simple instrumentation. We can use Foster resonance energy transfer sensors that are genetically encoded. These sensors are more sensitive and accurate. In this methodology, we need to develop a sensor that is suitable for the present study. Genetic engineering techniques must be considered so that the desired tissue will be expressed. We can measure and analyze the ions and metabolites via optical imaging techniques. Transcript profile data is available in high resolutions and this can be classified as an important advancement (Sozzani et al., 2012).

We can use functional genomics to detect the signaling pathways that are present in different cells. In the analysis of metabolic flux, the 2D HSQC or heteronuclear single quantum correlation can be considered as an important tool. In this technique, we can measure the isotopes. We can demarcate the glucosyl units that are present in the protein hydrolysate and starch hydrolysate of the biomass. We can identify the compartmentalization process in the plants with this technique. We can use NMR to illustrate the metabolic fluxes and biosynthetic pathways. This can be quantitively analyzed using isotope isomers or isotopomers. We can radiocarbon as a tracer compound. The carbon fluxes that are present in the metabolic networks can be detected and characterized. The labeling patterns that are present in the metabolites can be studied with the aid of radiocarbon labeling techniques. The NMR technique can provide information regarding isotopomers and carbon enrichments. The distribution of the isotopomers can be determined by analyzing the coupling patterns of ^{13}C and ^{13}C of the metabolites. There are many applications of the 13C - NMR technique in metabolic engineering studies. The flux maps were generated for transgenics and plant metabolic phenotypes.

Mass spectrometry can be classified as an accurate, reliable, and sensitive technique that can be used to estimate the labeling of the isotopes. This technique can be used to produce flux maps in metabolomics. It is always important to map the fluxes that occur in the cellular environment. This is possible with metabolic flux analysis or MFA. This can be used in the identification of the pathways. With this technique, we can illustrate the network regulation process that generally occurs in the biological system. We can use both NMR and MS to quantify the isotope isomers that were linked with the biological chemicals. NMR contributed much to enhance the sensitivity of MS. The isotope labeling experiments or ILEs can be classified as an important technique in analyzing metabolic fluxes. It is important to know that these techniques need mathematical modeling.

It is obvious that the metabolic fluxes determine the carbon traffic rates that are associated with the metabolic pathways and these are important in the physiology studies in plants. The process of quantifying the fluxes in the case of bulk metabolic networks can be problematic. We can measure a small number of fluxes directly. The cyclic pathways need to be measured indirectly. The intracellular fluxes can be indirectly quantified with the aid of isotopic MFA. The evaluation of flux needs models so that we can simulate the isotopomers in the flux. In metabolic databases like MetaCyc, the atomic arrangements and metabolic pathways have been documented. The design of the optimal isotope label can be determined by factors like substrates, instruments, true fluxes, and network structures.

The isotope labeling experiments can be important in evaluating the metabolic networks and their performance. Most of the metabolic databases are publicly available. Some of the open-source metabolic networks are KEGG PATHWAY and MetaCyc. These network databases can be used in the analysis of the metabolites. KEGG PATHWAY can be used to study the metabolism of most plant species. In this database, the pathway maps will be created manually. There will be an assignment of the enzyme orthologs. We can easily browse this database and compare the metabolites of several species. We can also estimate the pathways that are novel. We can analyze the expression of the genes with the available data. We can use the data present in this database with desktop tools. KegArray can be used in this process.

Another important database is MetaCyc. This database has data on the metabolism of almost all living systems. The pathway maps can be produced manually. The enzymes will be assigned for these pathways. We can search over the internet and can buildpathway maps. The construction of the pathway can be user-driven. We can use this database with desktop tools like Pathway. PlantCyc is another database that can be used to understand the metabolism of all the plant species. The pathways can be produced manually and enzyme orthologs will be assigned. We can browse for this database and can construct a pathway map. The construction of the pathways will be user-defined. Reactome is another important database that is available on the web. This Reactome database can be used to analyze the metabolic patterns of *Arabidopsis* . The curation of the maps can be based on KEGG and AraCyc. The database can be accessed over the internet. The Sol Genomics network can be used in the case of metabolomics in Solanaceae species. The putative enzymes will be assigned with the aid of MetaCyc pathway maps. We can access this database through the internet. The pathway construction will be user-defined. MetaCrop database can be used to understand the primary metabolism of the crops. The pathways will be manually curated with the aid of enzymes and maps. We can browse this database via the internet.

In the case of the KEGG PATHWAY, many reaction sequences will be fused and represented as maps. These maps can be called as KEGG reference pathway maps. The maps for a specific organism can be searched with the genome of interest. The organism-specific maps are available for poplar, maize, rice, and *Arabidopsis* species. Initially, these maps were constructed for animal and bacterial metabolism. These maps were also generated to illustrate the plant hormones and secondary metabolites (Kanehisa et al., 2011). MetaCyc database was developed to compare and understand the pathway variants and specific pathways. The IUBMB or International Union of Biochemistry and Molecular biology on the Nomenclature and Classification of Enzyme list was used to demarcate the enzymes. According to this system, there will be a four-digit code. The enzyme data can be stored in the form of databases and this data can be retrieved through a web application called as ExplorEnz. This data can be downloaded from the website in different forms (IUBMB). We can use the enzyme list provided by IUBMB in order to trace the pathways.

There are many crop genomes that were sequenced. This mass sequencing can help the researchers to understand the actual genetic data of the organisms. The complete genome sequencing process can help in understanding the compartmental organization of several organisms. The genome sequencing process was initially started with bacteria, yeast and others. The location of proteins will be inside the subcellular compartments that can hinder the annotation of these models. Most of the eukaryotic cells have a plasma membrane and they comprise all the structural components that are needed to carry on the metabolic pathways. The observation of the proteins in the organelles can be based on the signals (Jolley et al., 2010).

We can predict the presence of the targeting signals in the premature proteins via software that are available online. Most of these programs are based on algorithms. These algorithms are built upon data obtained from several species. The translocation of the proteins into the thylakoid membranes can be based on diverse transport mechanisms (Gutensohn et al., 2006). We can retrieve the candidate sequences via keyword searches, automated annotation systems and sequence conservation methods. The search results can be confined via elimination of redundancy, comparing sequence and counter checking. We need to validate the gene models via inspection of untranslated regions. We need to check the characteristics of introns and start/stop codons. Then, we can predict the localization of the proteins in the intracellular spaces. This can be done through experimental verification and predictive models. Finally, we can draw certain conclusions and make a hypothesis. In order to study the plant metabolisms, we need to frame a genome-scale model or GSM. This model can be classified as a silico metabolic model. In this model, we can analyze thousands of chemical reactions that are present in the organism, tissues, and cells. We can use GSM along with FBA to understand the metabolic fluxes. The GSM of any organism can be classified as an Insilco model that comprises of main metabolic reactions. GSM can help us understand the complete setup of organisms. There will be many reactions that can be present in GSM. The prediction and measurement of the carbon traffic can be possible with the quantitative analysis of the GSM models. We can quantify the GSMs with the aid of FBA or flux balance analysis. In this FBA technique, we can do quadratic optimization and linear optimization along with mass balancing. FBA can explain the flux distribution for many metabolic reactions.

The main reasons behind the unavailability of GSMs are the presence of secondary metabolite pathways and the high rate of compartmentalization. Compartmentalization can be considered as an important feature in plant metabolism. AraGEM model includes vacuoles, peroxisomes, plastids, mitochondria, and cytoplasm. In order to construct the GSM models, we need to collect a lot of information about the biological reactions that exist in organisms. The fields like transcriptomics, proteomics and metabolomics can be used to understand the changes in the protein concentrations and metabolites. We can even understand the changes that exist in gene expression. The structural and biochemical complexity can be a hurdle while engineering plants. In plant metabolic engineering, we can use many metabolic pathways of an organism so that we can easily understand them. The metabolic pathways should be translated into an appropriate mathematical model so that we can perform MFA. For many years, there were studies that focused on gene-metabolism relations and pathways. These studies were important to understand the secondary metabolisms in the plants. The metabolic profiling data and large-scale transcriptional data were combined to understand the genome sequence. Latest technologies paved the path to the advancement of our understanding of metabolites. Most of the metabolites are considered as products that were derived from the enzyme-mediated reactions. These products occur in living systems.

Chapter 7

Nutrient Deficiencies

CONTENTS

7.1 Introduction 138
7.2 Maize 139
7.3 Rice 143
7.4 Sorghum 146
7.5 Genome Sequencing 151
7.6 Phosphorus Uptake 153
7.7 Improved Crop Quality 154

7.1 INTRODUCTION

The challenge in the agriculture sector comes from its nutrient status and this will affect productivity all over the world. It is important to enhance crop production as there is a growing demand for food grains. It is important to increase the croplands. The production yield should be enhanced and must be part of intensive cultivation. Poor management of the nutrients and reduced soil fertility can be considered as the main problems in crop production.

Figure 7.1: Plant nutrient deficiencies.

Source:http://newcastlegardens.com/2017/07/nutrient-deficiencies-recommended-solutions/

The plants need balanced nutrition so that they can grow completely and give us yield. The soils can be considered as storehouses of many nutrients. Agricultural sustainability, soil fertility and plant growth are linked with each other. Plant growth can be associated with nutrients, water, carbon dioxide and solar energy. Crops suffer from unsatisfactory environments. The root penetration problems, shallow soil depth, lack of oxygen and carbon dioxide and others can be classified as problems that can determine the crop yields. The nutrient starvation results due to reduced concentration of copper, cobalt, manganese, molybdenum, boron, zinc, iron, sulfur, and other compounds. The toxic concentrations of chlorine, sodium, boron, copper, manganese, and iron can limit the crop production capacity. Increased application of fertilizers and manures can also enhance the crops for some time and then growth will be limited.

Most of the nutrient deficiencies in the plant species can be noticed as patches. These patches are common in the field conditions. They can appear due to nematodes that attack roots in some cases. There will be ambiguity between these symptoms. The root rot causing pathogen can also present patches in the field conditions. These symptoms are often confused with the symptoms presented by the nutritional disorders. These symptoms are more visible in the plants that suffer from N, P, K, Mg deficiencies. In case of heavy rains, there will be a loss of nutrients due to surface runoff. This will produce nitrogen deficiency. It is important to know that zinc deficiency is prevalent in most of the soils that suffer from heavy rains and lose their fertility. Excess application of K fertilizers can produce Mg deficiency in the crop plants. The liming of the acid soils can initiate iron deficiency. If the pH is less than 6.5, there will be Mo deficiency. In case the pH value is 7.5, there will be iron deficiency. We can observe Mg, Ca, K and P deficiency in soils that have a pH value of 6. Waterlogged conditions can produce P deficiency.

Most of the symptoms associated with nutrient deficiency can occur in a definite pattern. Due to Mn deficiency, there will be white specks in the maize crops and this is evident on the interveinal leaf areas. Zinc deficiency in the maize crops can be identified via white bands in the midrib margins of the leaves. Iron deficiency can cause interveinal chlorosis.

7.2 MAIZE

Maize can be more sensitive to N deficiency. Most of the symptoms linked with nutrient deficiencies can start with mild conditions. In N deficient soils, the leaves will appear yellow and there will be stunted growth. The size of the ears will be small with limited grains. The kernel size will be minimum and so the yields will be less. It is obvious that N is a mobile element. The N element can get transported from older leaves to the younger ones. The N deficiency primarily appears in the older leaves. If the plants suffer from N deficiency at their younger age, the entire plant will appear yellow. The younger leaves stay green. A V-shaped pattern can be observed as far as yellow chlorosis is concerned. If there is mild N deficiency, then the entire plant will appear light or pale green. There will be an appearance of green chlorosis in the older leaves if the deficiency becomes severe. We can observe necrosis at the tip of the leaves and this necrosis will reach the base of the leaf. The leaves that are affected will be brown and they soon die due to the absence of photosynthesis.

The soils that support less organic matter cannot withhold the N concentrations. The sandy soils with a light texture can get leached faster due to heavy precipitation. The leaching process can be due to irrigation also. The intensive cropping practices can cause the soils to become weaker and may not hold the nitrogen molecules. In the case of waterlogged conditions, the N shortage will be prevalent. If the pH of the soils falls below 6, then we can estimate that the soils are not having sufficient N molecules. If pH is greater than 8, then N deficiency can be imminent. It is always important to follow integrated nutrient management so that the crops can get the nutrients as they grow. It is always important to measure the amount of N that is present in the soils. We can use nitrogenous fertilizers, bio-fertilizers, and organic manures. We can dress the soils with urea doses. We can apply 2% urea solution on the leaf surfaces. This application must be repeated twice a month. It is better to use legume crops in the crop rotation process.

In the case of P deficiency, we can observe stunted growth with dark green leaves. The size and numbers of stomata will be reduced. The growth of roots will be diminished. We can observe small ears with poor grain yield. P is also a mobile element. It can move from old leaves to younger ones. The symptoms appear in older leaves. The old leaves will appear as dark green. Occasionally we can observe bluish-green leaves. Most of the leaves appear burnt and they eventually die. In the developmental stages, we can observe mild deficiency symptoms. The symptoms appear in the older leaves. P deficiency can occur due to reduced organic matter in the soils. The calcareous and alkaline soils may not support P effectively. The intensive cropping practices can deprive the soils of the P residues. In case topsoils are removed, then the P deficiency will occur. If the soil pH is less than 6 or greater than 7.5, P deficiency can occur in the soils. P deficiency can be managed by integrated nutrient management. We need to check the soils prior to sowing so that the phosphates that are present in the soil can be evaluated and analyzed. We need to use phosphatic fertilizers, phosphate solubilizing microbial cultures, and organic manures to regulate the P concentration in the soils. It is important to add ammonium phosphates in the water especially with irrigation water.

In the case of K deficiency, there will be a shortening of the internodes. The plants will appear as dwarfs. The foliage will appear dark green. The plants that suffer from K can have reduced ears and they appear pointy. The tip of these ears will be underdeveloped in most cases. K can travel from old leaves to the younger ones. The symptoms appear first in the old leaves. The young leaves can be green. Generally,pale-yellow chlorosis appears on the leaves. After chlorosis, there will be necrosis. Marginal chlorosis will be present in the older leaves. We can even find red strips at the leaf sheaths and lower portions of the stems. In the case of mild K deficiency, the plants will exhibit stunted growth, pale green foliage, and shortened stems. In extreme cases, the leaves will get burned and eventually die. The P deficiency can be caused due to low K in the soils. The K from the parent materials will not be supplied to the soils. The K ions will be leached away due to excess irrigation and heavy precipitation. The soils may have minimal organic matter. If the bicarbonate concentration is high in the irrigation water, then there will K loss. The acid soils with pH 6 cannot hold K ions. The soils must be analyzed before sowing. The soils that are deficient with K must be reclaimed. We need to add organic manures prior to sowing. We need to add potassium chloride, potassium sulfate and potassium nitrate to the soils well before the sowing procedure. We need to add these chemicals as foliar sprays.

The calcium deficiency in the maize plants can reduce the crop yields. The calcium deficient crops will be stunted and there will be disturbed tassels and small ears. The calcium ions can be classified as immobile elements and they cannot migrate from old leaves to younger leaves. There will be yellowish inter-veinal lesions. Bull-whip like structures gets developed due to Ca deficiency. Ragged appearance can be prevalent as there will be hole formation in the leaves. Ladder like appearance is observed in the Ca deficient plants. The younger leaves may not get separated from the whorls in many cases. The acid sandy soils can lose calcium ions easily due to leaching. The sodic soils are more vulnerable. The mulch soils and acid peat can have low calcium concentrations. The soils that have a pH less than 6.5 and more than 8.5 can be unsuitable to hold calcium ions. As a part of integrated nutrient management, we need to analyze the calcium content in the soils. We need to add calcium chloride, calcium nitrate, and calcium sulfate to the soils via fertilizers. In the case of acidic soils, we need to add limestone or lime. Dolomite can be added to the soils. Calcium sulfate can be added to the crops in the form of foliar sprays.

The magnesium deficiency in the maize plants can reduce crop yields. The foliage will be pale green and occasionally there will be brown leaves at the bottom. The kernels will be small. The magnesium molecules can be classified as a mobile element as they can transfer from old to young leaves. Most of these symptoms appear in the old leaves. We can observe pale yellow inter-veinal chlorosis. The tip and base of the leaves will be affected. There will be brown strips in the older leaves. There will be the formation of necrotic strips in the leaves. The older leaves die soon. The Mg deficiency mostly occurs in the acid sandy soils. These soils are more vulnerable to rainfall leaching. The muck soils and peat can exhibit low Mg concentrations. Some soils that have high K and Ca concentrations can exhibit low Mg concentrations. The soils that were formed due to poor weathering of the parent material can exhibit Mg deficiency. The soils that have pH values less than 6.5 and greater than 8.5 can lose Mg ions faster than other soils. As part of integrated nutrient management, the soils must be analyzed for Mg concentration. We need to add magnesium chloride or magnesium sulfate to the soils before sowing. We can add magnesium nitrate, magnesium chloride, or magnesium sulfates to the soils via irrigation water. We can add dolomite to the acidic soils. Dolomite can be defined as a mixture of magnesium carbonate and calcium carbonate. It is always important to reclaim the acid and basic soils so that the magnesium ions can withstand. Sulfur deficiency in the maize crops can produce more damage to the crops. Due to S deficiency, the foliage will be yellow in color.

Boron can be classified as an essential plant nutrient. It can get involved in hormone functioning, sugar translocation, cell elongation, and cell division. There will be growth-related defects if B is deficient in the plants. The growth of the pollen tubes can be based on B concentrations. If B is limited in the plant systems, the pollination process will be hindered. The growth of roots will be reduced. We can notice pale green foliage and stout stems in case of B deficiency. There will be bent cobs and small cobs. Due to B deficiency, there will be reduced ears with minimal grains. This element can be classified as immobile. It cannot reach the younger leaves from the older leaves. This is the reason why the older leaves will remain green. Most of these symptoms can be seen in the younger leaves. We can notice the puckering of leaves in the case of B deficiency. There will be transparent stripes and watery stripes in the leaves. The low moisture conditions and drought can bring about B deficiency in the crops. The drought conditions will tamper with the B absorption process. The acidic igneous rocks can weather and yield B deficient soils. The sandy soils can get leached due to heavy precipitation and, in some cases, excess irrigation also pushes the soil to lose B. Most of the alkaline soils that have lime cannot store b molecules. The soils that have less organic matter cannot store B molecules. The acidic and alkaline soils may not hold B molecules. The soils must be analyzed prior to the sowing process. We need to apply chelated Boron, boric acid, and borax to enhance the B concentration in the soils. The organic manures must be added to the soils.

7.3 RICE

The N deficient rice crops can be stunted and they have yellowish-green leaves. The number of grains will be reduced. The nitrogen element can be classified as a mobile element. The N molecules can move from older leaves to younger leaves. This is the reason why the symptoms appear in the older leaves. The entire rice crop will appear yellowish-green. The older leaves will be pale yellow. Most of the yellow chlorotic leaves can turn brown and ultimately die. The soils that support low organic matter may not hold the N molecules. Insufficient application of N fertilizers can reduce the N concentration in the soils. Due to intensive cropping, most of the soil nutrients will be lost. Heavy precipitation can push the soils towards leaching and so N molecules will be lost. The submerged conditions can cause N deficiency. The drying seasons whilst cropping can also result in the N deficiency. The reduced biological nitrogen fixation can also hinder the N accumulation in the soils. The acid and alkaline soils cannot support the N contents in the soil. As usual, we need to analyze the soils prior to the sowing period. We can add nitrogenous fertilizers, bio-fertilizers, and organic manures to enhance the N concentrations. We can add urea super granules and S coated urea to enhance the N content in the croplands. The slow release mode of the fertilizers can help much in this regard. We need to split doses of urea application so that the crop plants can use most of them. Foliar sprays can be used when the crops are already in their growing stages. We need to use legume crops in the crop rotation process so that biological nitrogen fixation can be achieved. The Phosphorus limited rice plants can appear dark green but there will be erect leaves. Most of these plants exhibit stunted growth.

The panicles will be reduced and so the grains. The application of N can be useless unless P is used. The P molecules can be classified as mobile. They can migrate from the older leaves to the younger ones. The leaves will be dark green. We can notice purple and red colors in the older leaves. The leaf margins will be purple in color and this coloration will extend to the entire leaf. Most of the older leaves eventually turn brown and then die. Due to intensive cropping, the soils will not hold this element. Excess liming in the soils can also result in the immobilization of this element. The soils that have coarse texture can have low organic matter and so the P-reserved will be reduced. The P fixation capacity will be high in the acid upland soils and clayey soils. Even highly weathered soils can also have P deficiency. The sodic soils, saline soils, and calcareous soils can exhibit P deficiency. If we use insufficient phosphatic fertilizer and excess nitrogenous fertilizer, then there will be P deficiency in the soils. The acidic and basic soils may not hold P molecules. It is important to test the soils before sowing season. We need to add phosphatic fertilizers, phosphate solubilizing microbes, and organic manures to the soils. Ammonium phosphates can be added to the soils.

The rice crops that are facing K deficiency can be severe. The genetically engineered or hybrid plants are more vulnerable to K deficiency. K molecules can migrate from the older leaves to the younger ones. The symptoms of this deficiency can appear in the older leaves. There will be dark rust spots that are brown. The soils that were formed due to improper weathering can make the soils K deficient. If the soils have less organic matter, then they may not hold K ions. Increased use of phosphatic fertilizers and nitrogenous fertilizers can cause the limited absorption of K. Sowing of the rice seeds in the initial growth stages can cause this deficiency. The hybrid rice cultivars can be more susceptible. As part of integrated nutrient management, we need to analyze the soil for available K. It is important to add organic manures prior to the planting process. We need to add potassium chloride, potassium sulfate, and potassium nitrate. The K solution can be applied in split doses of three.

S deficiency is common in rice plants. Due to S deficiency, there will be stunted growth with pale yellow leaves. The grains will be reduced and there will be prolonged crop maturity. The S molecules can be classified as less mobile. The symptoms first appear in the young leaves. In this case, the younger leaves can be yellow and the older leaves will be green. Necrosis may appear in the case of prolonged S deficiency. This deficiency is most common in the case of less organic matter in the soils. The sandy soils that are light textured can be subjected to leaching due to heavy precipitation. Due to intensive cropping, the S deficiency will be prevalent. The weathering from the parent rock may derive S deficient soils. Most of the acidic soils lose S molecules. As usual, it is always important to test the soils for available sulfate molecules. We need to add gypsum and elemental sulfur to compensate for the S loss. We can add potassium sulfate, magnesium sulfate and ammonium sulfate to the soils.

Iron deficiency is another prominent problem in rice crops. Iron deficiency is common in the upland soils, aerobic soils, and high pH soils. The iron deficiency may not be seen in the lowland soils that were flooded. The concentration of iron molecules will be increased due to flooding. Mostly, iron can be considered as an immobile element in plant systems. The symptoms of iron deficiency may occur first in the younger leaves. There are no noticeable symptoms in the older leaves. There will be a fading of leaves. In no time, the plants may return to their normal growth conditions. Chlorosis may appear in case of severe deficiency. Most of the veins can remain green. When the symptoms start to increase, we can find yellowish-white chlorosis. The new leaves can fall off in case of severe deficiency. The leaves become papery and they die due to necrosis. The upland soils can exhibit less soluble iron in its ferrous form. Most of the soils that have less organic matter may not support the iron molecules. The bicarbonates can lower the solubility rates of the iron molecules and hence lowered uptake. This can be observed in the calcareous soils and alkaline soils. The iron phosphates may not be available to the plants. Excess use of phosphatic fertilizers can decrease iron uptake. It is important to analyze the soils prior to sowing. The soils that are Fe deficient must be reclaimed. We need to add organic manures. We can use ammonium sulfate-based fertilizers to enhance the availability of iron molecules. We can use ferrous sulfate in its solid state.

Rice can be most susceptible to zinc deficiency. This Zn deficiency can damage the entire crop. This deficiency can cause sterility in the spikelet, reduced root growth and stunted plant growth. We can observe brown batches on the leaves. The whole plant will have a blotch-like appearance. The sunlight will enhance the damage. In some cases, the entire leaf dies. The zinc deficiency can be seen in leached sandy soils. In the case of leveed soils, the surface soils will have more zinc content. Excess use of phosphate fertilizers can reduce the production of zinc phosphates. The soils that have high bicarbonates can exhibit Zn deficiency. The zinc content will be decreased with an increase in P, Mn, Cu, Mg, Ca, and iron availability. The soils that have more liming can exhibit Zn deficiency. The acidic soils and basic soils can exhibit Zn deficiency. The soils must be tested for Zn deficiency. The alkaline soils should be reclaimed. It is important to know that zinc fertilizers should not be mixed with P-based fertilizers.

Boron can be considered an essential element in plant growth. Most of the growth-related mechanisms are linked with sugar translocation, cell elongation, and cell division. Boron can be considered as an immobile element in plant systems. The deficiency symptoms will appear in the younger leaves. The old leaves will remain healthy. The leaves will have a white tip. After some time, the leaf starts rolling. The drought conditions can hinder B absorption. The limited soil moisture can reduce B absorption conditions. The soils that are weathered from the parent rock may contain less B. The freshwater sediments and acid igneous rocks can have limited B concentrations. The sandy soils can be leached easily due to heavy rains. Excess irrigation can also push the soils to lose B content. The basic soils will have more lime. The acidic and basic soils may not hold B content.

7.4 SORGHUM

Sorghum plants can be considered a fast-growing species. Most of the sorghum species are sensitive to N deficiency. There will be reduced grain yield and fodder production. N deficient plants exhibit stunted growth. The leaves will be yellow. The nitrogen molecules in the sorghum plants can be classified as mobile. The N molecules can move from old leaves to younger ones. The symptoms pertaining to N deficiency can appear in the older leaves. There will be chlorosis that can be yellow. We can find a V-shaped pattern in the chlorosis situation.

The soils that have less organic matter may not support the nitrogen concentrations. The soils that have light texture can be leached easily due to the heavy rainfall, Intensive cropping can also affect the availability of this element, most of the soils that are in waterlogged conditions may face this problem. The extremely acidic and basic soils may lose this element. As usual, it is always important to get the soils tested before sowing. We need to add nitrogenous fertilizers, bio-fertilizers, and organic manures. We need to add urea in split doses and foliar sprays may help in this regard. It is important to add legume crops.

Sorghum crops face phosphorus deficiency. The sorghum plants can be classified as a better indicator of P deficiency. The P deficient plants can appear as stunted and there will be dark green leaves that are spindly. Due to P deficiency, there will be slow crop maturity. Mostly the younger leaves will remain normal. The older leaves can appear as bluish-green. The purpling of the entire plant is common in the case of P deficiency. The plants eventually die due to P deficiency. The soils that can store organic matter may not hold P. The calcareous soils and alkaline soils may not withstand P molecules. The topsoil may be removed due to erosion caused by rains. Extremely basic or acidic soils cannot sustain P for a while. To manage the nutrients in the soils, we need to examine the P molecules in the soils. We can add phosphatic fertilizers and organic manures. The microbial cultures that are P solubilizing can be used. Ammonium phosphate can be used to restore the P molecules.

K deficiency can be a main problem in sorghum plants. Due to K deficiency, there will be shorter internodes. This will result in the production of dwarf plants. The foliage will be dark green. There will be minute heads due to K deficiency. The number of grains per plant will be reduced. The K molecules can move from older leaves to younger leaves. Most of the symptoms linked with K deficiency can start with the appearance of chlorosis and necrosis. These symptoms mostly appear on the older leaves. The parent rock may have less K concentration so on weathering the soils produced will also have less K. The soils that are light textured can hold limited K. This is because irrigation and heavy rainfall via runoff can enhance K loss. The limited proportion of organic matter cannot support K ions. The acidic, basic, and acute sodic soils may not store K ions. The bicarbonate concentration may not support the K ions. The acidic soils may not support K. It is always important to test the soils. The saline soils, alkaline soils, and other unsuitable soils should be reclaimed. We need to add organic manures well before the sowing periods. We can add potassium chloride, potassium sulphate, and potassium nitrate to the soils to restore the K ions in the soils. The foliar sprays will be useful.

Calcium deficiency is most observed in the sorghum crops. There will be ragged foliage and distorted foliage. Most of the plants with this deficiency die before reaching maturity. The calcium element can be classified as immobile. This element may not reach the younger leaves. Most of the older leaves will remain healthy and green. There will be yellowish lesions on the leaves. Most of the leaves will exhibit a bull whip-like appearance. There will be holes or apertures in the lamina of the leaves. Due to these malformed and torn leaves, there will be a ragged appearance. The leaves may not get separated from the whorls. Most of the sandy soils that are acidic will lose calcium ions due to heavy rains. The sodic soils with sodium ions cannot hold Ca ions. The soils that bear low exchangeable calcium and soluble aluminum may not withstand the calcium ions. The muck soils and acid peats can have minimal calcium ions. The extremely acidic and basic soils cannot hold calcium ions for long periods. It is always important to test the soils for calcium deficiency. We need to use fertilizers that have more calcium. The calcium content in the soils can be enhanced by applying calcium chloride, calcium nitrate, calcium sulfate along with other fertilizers. In the case of acidic soils, we can use limestone and lime as calcium supplements. We can supply the calcium ions in the form of foliar sprays. (Clark et al., 2008).

The sorghum plants that are S-deficient can exhibit yellow foliage and in some cases, we can observe pale green foliage. There will be minimal crop yields and a smaller number of grains. The symptoms of S deficiency should not be confused with the N deficiency symptoms. The older leaves appear as dark green. The S deficiency can result in the necrosis of the leaves. There will be a change in the leaf color. The sulfur compounds can be classified as immobile elements. The younger leaves suffer from S deficiency first. We can observe the pale young leaves. The soils that have less organic matter can hold less S. The sandy soils with light texture can get leached soon and it may also be due to excess irrigation. The intensive cropping patterns can also increase S loss. If the parental material is S deficient then the soils that are weathered from the parent rock can suffer from the same deficiency. The acid soils with low pH can also suffer from S deficiency. We can compensate for the S loss by applying gypsum and elemental sulfur. We can use potassium sulfate, magnesium sulfate and ammonium sulfate. The problematic soils must be reclaimed. (Grundon et al., 1987).

Iron deficiency can be observed in sorghum plants. We can observe chlorosis, especially in the basic calcareous soils. Iron molecules can be considered as immobile elements. The symptoms are visible in the younger leaves. Most of the older leaves will remain unaffected (Follett et al., 1971). In severe iron-deficient conditions, there will be burns on the plants. Most of the sandy soils will have limited iron concentrations. The calcareous soils will exhibit limited soluble iron. The muck and peat soils have an organic matter that can absorb iron molecules and so the iron concentration in the soil solution will be reduced and might not be available to the plants. The acidic soils that may have other elements may not be suitable for the plants so that they can absorb iron molecules. The waterlogged soils may not hold the iron molecules. Extremely basic soils may not withstand the iron molecules. It is important to perform soil tests prior to the sowing season. We can add inorganic soluble salts like iron chlorides and iron sulfates to the soils. We can use iron chelates. We can add FeEDTA that can chelate the iron molecules.

The sorghum crop will be more susceptible to zinc deficiency. Due to Zn deficiency, there will be a shortening of the internodes. Due to this, there will be stunted growth. As a result, we can expect a poor yield. Zinc ions may not transfer to the younger leaves. We can observe yellow streaks and bands on the base of the leaves. The leaf margins and midrib will be green. The sandy soils may leach and so the zinc molecules will be minimal. Most of the subsoil will be exposed due to cultivation. The surface zinc availability will be high. The cold weather conditions will also hinder the availability of Zn ions. The application of phosphate fertilizers can reduce zinc absorption rates. The extremely acidic and basic soils may not hold the zinc molecules. The problematic soils should be reclaimed. We need to add organic manures prior to sowing. The zinc fertilizers must not be mixed with the phosphate fertilizers (Ohki et al., 1984).

Manganese deficiency can result in stunted plant growth. The leaves will turn yellow. The manganese molecules are classified as immobile elements. The older leaves will not be affected. There will be interveinal chlorosis. We can detect white flecks due to Zn deficiency. The sandy soils that have light texture may lose Zn due to rainfall. Excess irrigation can also remove zinc ions from the soils. The alkaline soils and calcareous soils cannot hold zinc. The problematic soils should be reclaimed immediately. We need to add organic manures. Magnesium salts like magnesium sulfates should be added to the soil. The pearl millet crop can be susceptible to N deficiency. Due to N deficiency, there will be stunted growth. We can notice spindly stems. The leaves may become yellow and pale green.

The characteristics of the nutrients and microbes that regulate crop growth can be considered as important. These characteristics were studied using a split-bock experiment. This experiment was devised with diverse fertilizer treatments. The major nutrients like nitrogen, potassium, and phosphorus were used in this experiment. In the South Asian countries, the rice-wheat cropping system is important and this is since rice and wheat crops are required in huge quantities as most of the population rely on these food products. These crops suffer from manganese, boron, and zinc deficiencies. The wheat and rice crops must be grown in crop rotation mode so that the soil fertility will be intact. The imbalances in the organic matter and fertilizer use can damage the whole soil structure and this will reduce the metabolism of both crops. The micronutrients can be considered as important elements that can control membrane integrity, enzyme activation, photosynthesis, and chlorophyll production. These micronutrients can be absorbed via rhizosphere and it is based on the physicochemical nature of the soils. Some of the important soil properties are soil moisture, organic matter, and pH. The tendency of the plants to absorb the nutrients can be based on many factors. Some of them are root exudates, transport kinetics, root hairs, and root architecture. The crops should be managed appropriately so that the micronutrients can be used by the plants easily. The rice-wheat cropping systems can be important in countries like Nepal, Bangladesh, and India. The productivity and sustainability of the cropping systems are being damaged due to poor micronutrient management and damaged soil health. Due to these factors, there will be reduced grain quality and so the yield. The poorly managed agricultural practices can result in micronutrient deficiency in the Indo-Gangetic plains.

The intercropping technique can ease the cropping index and so the utilization rates. This technique can enhance yields at a constant rate. The intercropping technique can be used in sustainable agricultural development. In this technique, there will be two crops or more and they can survive for some time. In Africa and America, cassava is grown as an intercrop. In the case of Southern Africa and Eastern Africa, legume-cereal intercropping is a common practice. In most of the temperate zones, we can use this technique. In China, maize and pea, sunflower-soybean, maize-soybean, and wheat-maize are popular crops that can be grown as intercropping crops. The crops can use the resources effectively and there will be niche differentiation. The legumes can be intercropped with grasses so that the crops can easily absorb nitrogen. Crop intensification and intercropping can enhance the land utilization capacity in croplands and landraces. We can use the light environment effectively if more than two species are being grown. Most of the crop nutrient requirements are based on growth condition variabilities, photosynthetic rates, and morphology of the leaves primarily.

We can obtain the positive results of intercropping only if there is ample space and sufficient time. With minimal space and time, we can expect competition between the species and so there will be demand for all the nutrients and environmental conditions. It is always important to alter the spatial and temporal patterns if necessary, as the plant populations can get affected by these parameters. Whilst studying the nutrient use efficiencies, it is important to understand the utilization efficiency and nutrient absorption. There should be an established relationship between the utilization efficiency and land equivalent ratio along with the rate of nutrient absorption capabilities.

7.5 GENOME SEQUENCING

The central dogma recommends that most of the biological processes must be derived from genomic DNA (Figure 7.1).

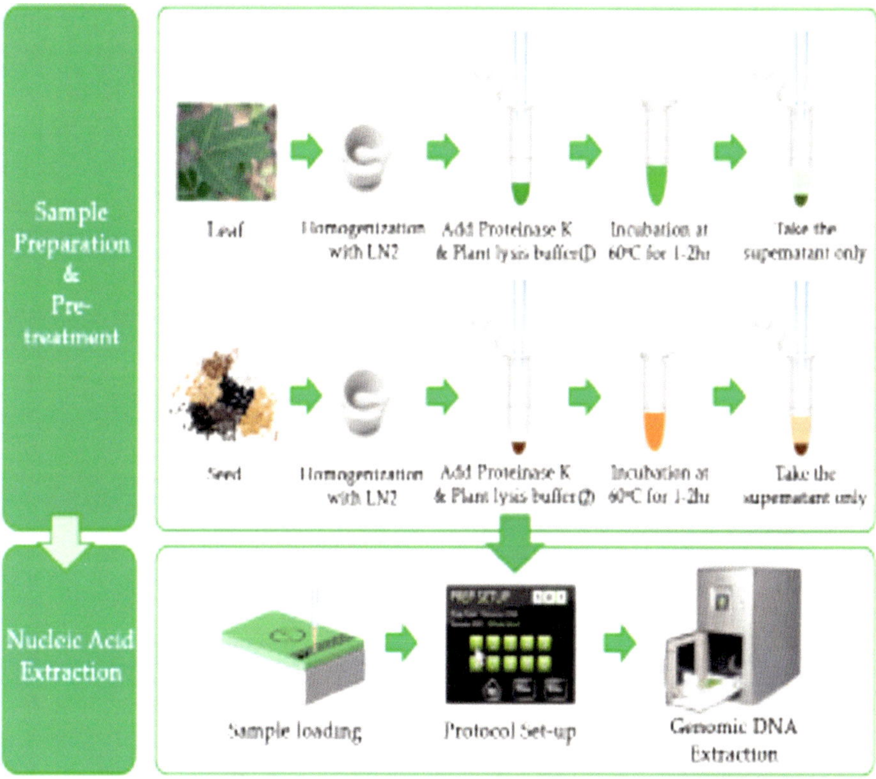

Figure 7.2: Genomic DNA.

Source:https://us.bioneer.com/products/accuprep/ExiProgenPlantGenomicDNAKit-technical.aspx

The genome can be classified as an important entity in life forms. It is important to know the genome sequence of organisms under study. It was understood that genome-wide studies help in understanding biological systems. Genome-wide sequencing is very important in understanding almost all the organisms. The genome-wide sequencing took several years with the Sanger method. Next-generation sequencing or NGS technologies (Figure 7.3) have enhanced our understanding of the genomes like never. Genome sequencing is quite economical than conventional sequencing techniques. The NGS based studies are being used in understanding many genomes.

Pacific Biosciences or PacBio is a single molecule real-time sequencing method that can be used in genome sequencing studies. This technique can be used in the genome sequencing of several crop species. The International Rice Genome Sequencing Project or IRGSP has produced many methods that are still being used by the research communities worldwide. The draft genomes of rice were made available before the IRGSP. The two species of rice are japonica and indica. IRGSP has made available the entire genome of japonica rice cultivar in the year 2004 (IRGSP 2005). In this project, clone by clone Sanger sequencing and map-based sequencing were used. The fosmid clones, PACs and BACs were sequenced and the genome assembly was made. The genome libraries were used to construct genetic maps. The rice genome was about 389 Mb. The annotation of the genomes can be important whilst knowing the information pertaining to the genomes in biological studies.

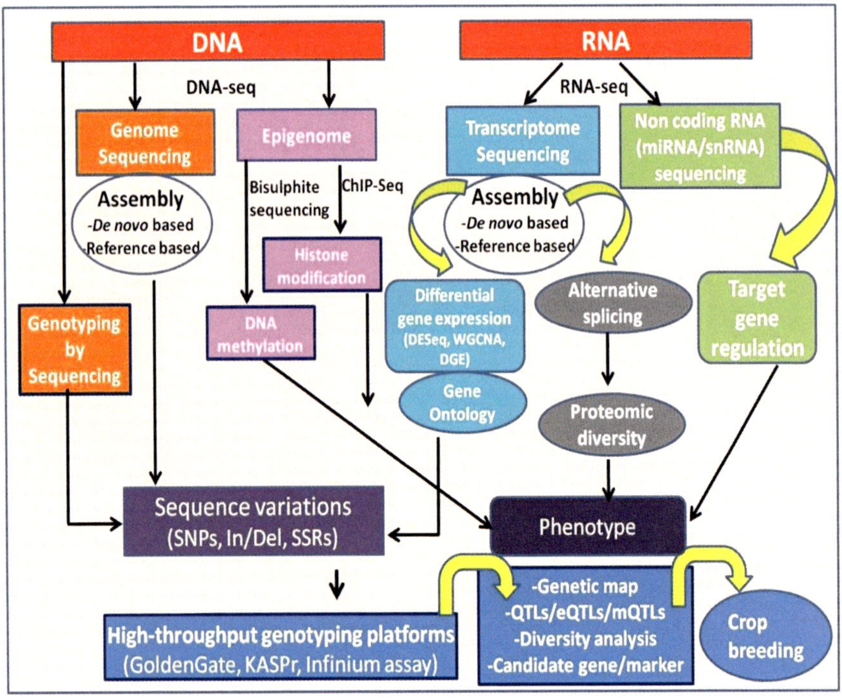

Figure 7.3: NGS technologies.

Source:https://www.intechopen.com/books/next-generation-sequencing-advances-applications-and-challenges/reaping-the-benefits-of-next-generation-sequencing-technologies-for-crop-improvement-solanaceae

The small RNAs can be considered as important components of the regulatory pathways of the gene. These small RNAs are classified as conserved in the case of eukaryotes. Artificial microRNAs and RNA interference or RNAi can be used to alter the expression of the gene. The RNAi was mostly studied in the insect cells and nematodes. The small RNAs can regulate most of the biological mechanisms of the plant systems. The small RNA can be about 20 to 30 nucleotides in length and most cases, we classify them as small interfering RNAs. This assumption and classification are based on the structure of the precursors and their processing. The miRNAs can be derived from the RNAs that are single-stranded and have loops. The small RNAs or siRNAs can be produced from the double-stranded RNAs. These molecules will be complementary. The small RNAs in rice are miRNAs, ta-siRNAs or phasiRNAs, nat-siRNAs, and hc-siRNAs. The precursor in the case of miRNAs is single-stranded RNAs and the other three have double-stranded RNAs as precursors. The profiles of small RNAs were analyzed in several plant species. Some portions of miRNAs were considered as conserved in the terrestrial plants. With this assumption, we can say that small RNA dependent regulation of the gene is mandatory for the survival of the plants. The siRNAs are abundant in the rice genome.

7.6 PHOSPHORUS UPTAKE

In the case of annual plants, the period of senescence can be important as it involves the transfer of C compounds and nutrients to the seeds and other organs of the plant body. Most of the soils are deficient in P and the plants need to store P in the seeds. The seeds will survive the P deficient conditions and may withstand competition from others. In the native ecosystems, the P levels in the soil are less whereas in the agricultural soils P is more. It is always important to maintain P levels in the soils so that the P molecules will be available for the crops. It is important to know that the loading of the P molecules in the grains is genetically conserved. Minimizing the number of P molecules that are being absorbed by the seeds can keep the soils enriched with P and this can be important in integrated nutrient management. Once the plants reach maturity, P can be present in the grains of cereals. The P can be isolated in the form of phytic acid. These molecules may not be useful for humans and so we can observe loads of these chemicals in the excreta. The animal manures are are excreta from the animals. Heavy addition of animal manures can enhance the P loads in lakes and rivers. The P molecules present in the grains can be from endogenous and exogenous sources. The endogenous sources are the vegetative parts of the plants that can absorb P molecules. The exogenous resources are the uptake of P molecules via roots from the soils. High P uptake in the rice crops was observed post the flowering stage. The P molecules that are absorbed by the plants for grain filling can reach the epidermis of the nucellar regions. In some cases, these molecules can also reach aleurone cells. The P molecules can be transported through xylem vessels.

7.7 IMPROVED CROP QUALITY

The systematic improvement of the quality and standard of life pushed the natural resources to their fullest extent. There is a clear demand for food resources with the quality intact. It is always important to consider the quality of food products as a function of demand and vice versa. Water stress can be considered an important factor that can limit crop production rates.

Nanotechnology is being used to enhance the quality of rice crops all over the world. The current challenge in the agriculture field is the decline in food quality and quantity. Nanotechnology is being used in the agri-crops to enhance yields. Nanotechnology research has become significantly important in agriculture due to climate change, demand for food and allied products along with population growth. Nano-carbon can be used as a coating agent in many fertilizers as it can drastically reduce water pollution. Nano-carbon can be used in the slow release of fertilizers. The nanoparticles and nanotechnology can be used as magic bullets. These particles can be loaded with nucleic acids, fertilizers, fungicides, and herbicides. These bullets can target the plant tissues. The required materials can be supplied to specific parts of the plant tissues.

Latest farming methods are improving the rural communities and the field ecosystems are being improved. With integrated nutrient management, we can enhance crop quality and nutrient reutilizing. An integrated farming system can enhance the yield and revenue of the crops. There will be enhanced soil and water index. There will be improved nutrient cycling. The chemical fertilizers can enhance the pollution rates in the soils and so the native organic matter will be damaged. It is important to discourage farmers from using chemical fertilizers. Ecological farming can help in improving the cropping systems all over the world. It can encourage farmers to use farmyard manure, green manures, oils and household wastes. The traditional farming systems can reduce environmental damage and simultaneously lower soil fertility. The physical and chemical factors in the environment are important in determining crop growth patterns. The chemical fertilizers can degrade the environment and may strain the ecosystem's safety. Anthropogenic interference can alter the native course of all the biogeochemical cycles. The biotic communities can be damaged. There will be a loss of biodiversity across various ecosystems. The traditional methods of agriculture needed heavy doses of agrochemicals that can pollute the ecosystems.

Chapter 8

Nanotechnology in Crop Nutrients Studies

CONTENTS

8.1 Introduction 158

8.2 Biomass and Yield 169

8.3 Nano Fertilizers 173

8.4 Nanopesticides 178

8.5 Applications of Nanotechnology 184

8.6 Management of Pests 186

8.7 Environmental Problems and Challenges 191

8.8 Indirect Measurement Techniques 194

8.9 Conclusion 196

8.1 INTRODUCTION

The nanomaterials and nanoparticles can be utilized in supplying nutrients to plants. The nanoparticles can be used as nano fertilizers. These nano fertilizers will have positive and negative effects. The nano fertilizers can affect the soils, plants, and soil biota. These effects can be based on several factors. Some of them are soil dynamics, plant species, and the inherent properties of the nanoparticles. Most of these nanoparticles will minimize the microbial communities of the soils. These nano fertilizers can enhance nutrient use efficiencies. The effect of these particles can be slow but effective. These particles can lower the nutrient losses in any agroecosystem. The farm productivity will be enhanced on the usage of these nanoparticles. These particles rejuvenate the soils. We can use natural zeolites in this case. Most of the nano-clays can be useful in modifying the soils. There are many applications of nanotechnology in agriculture and crop nutrient uptake. We can use nano-clays and zeolites to enhance soil quality. The plant nano-nutrition can deal with the use of nanonutrients in the plant metabolism, translocation, nutrient uptake and influence the elements in the rhizosphere regions. This branch of science deal with phytotoxicity and other plant nutrition stages. Nanotechnology deals with the substances that fall on a nanometer scale.

The agri-nanotechnology can be classified into phytonanotechnology, bio-nanotechnology, and zoo-nanotechnology (Figure 8.1). The phytonanotechnology deals with nanoparticles in plant production systems and plant sciences. This technology can enhance crop productivity via the usage of agrochemicals that are in the nanoscale range. They are classified as nanofertilizers and nanopesticides. We can enhance the soil quality through increased water holding capacity. For this, we can use hydrogels, nano-zeolites, and nanoclays. These particles or substances can activate plant growth. This growth can be due to increased nutrient uptake. We can use carbon nanotubes, titanium dioxide, and zinc oxides in this case. We can use remote sensing devices equipped with nanosensors to monitor agricultural growth. The nanotechnology applications in agriculture were promoted by most of the industries and universities. The ecotoxicology of these nanomaterials must be evaluated before their application in the field. The nanomaterials may cause pollution that may not be detected by our sophisticated instruments. The biosafety of the nanomaterials must be considered before using these particles. It is important to enhance efforts in nanomaterials as they give us enough benefits. Silver nanoparticles in the soils can be considered as an important issue in this case. The divalent cations and ionic strengths can be important to allow the aggregation of the silver nanoparticles in the soils. With an increase in pH in the soils, the bioavailability of the silver nanoparticles will be reduced. It is also important to know that there will be hetero-aggregation of the nanoparticles with the colloids in the oil. Due to this, the mobility rates of these nanoparticles will be minimized and hence may not be available for the plants. The CEC rates can be increased leading to the surfacing of the Ag nanoparticles.

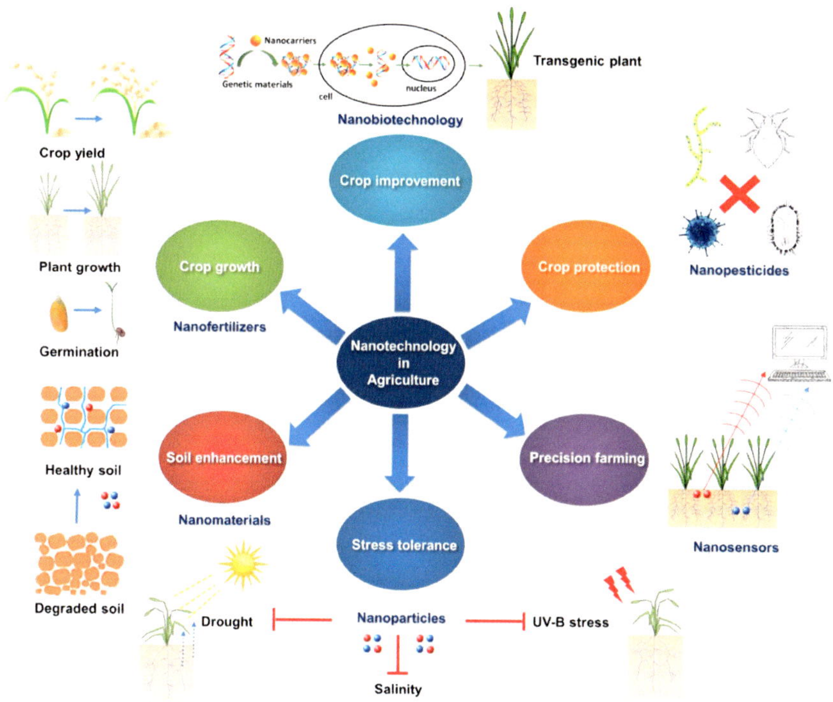

Figure 8.1: Nanotechnology in Agriculture.

Source: https://www.mdpi.com/1420-3049/24/14/2558/htm

The association between the agroecosystems and nanoparticles can differ drastically. The metal oxide nanoparticles can be homogenous and heterogenous whilst maintaining environmental dynamics. The toxicity, transformation, transport, and fate of the nanoparticles can be considered as important factors in agricultural nanotechnology. The pollutants that are present in the environment can change the course of the nanoparticles in the soil solution. The interaction between the metal nanoparticles and pollutants can determine the fate of nanocomposites in the soils. The presence of pollutants and contaminants can determine the bioavailability of the nanoparticles. Most of the pollutants can inflict synergistic toxic evasions on the plants.

The soil ecosystems and human life can be in jeopardy if the nanocomposite and nanomaterials are not used judiciously. The nano zero-valent iron or nZVI can induce ecotoxicity in several biomes. In rice crops, nano-CuO increases the proline contents. In maize plants, nano CuO decreased the succinate dehydrogenase activity. The nanomaterials can protect the plants from ROS or reactive oxygen species. These nanoparticles can increase the antioxidative enzymes. Some of the enzymes are peroxidase, catalase, and superoxide dismutase. The nanoparticles can reduce the oxidation stress in the plants. The nanoparticles can easily enter the plant tissues due to minute sizes. These particles can enter the plant tissues via water channels. These nanoparticles help in seed germination and growth. The surface area of the nanoparticles will be high and so the adsorption rates will be high. This will help us in the easy delivery of the nano-based agricultural products. We can encapsulate the nutrients in the nanoporous materials. We can coat these fertilizers with polymer films. The nutrients can be easily supplied to the plants with limited effort. The fertilizers that are present within the nano dimension formulations can be called nano fertilizers. They can be easily supplied to the plants without much rejection. Under controlled environmental conditions, these nano fertilizers can enhance the productivity of the crops in terms of growth and development. In tomato crops, cerium oxide nanoparticles were used to enhance the nutritional quality of the fruits. The application of nanomaterials and nanoparticles in the nutrition of the plants can be classified as plant nano-nutrition. The nano nutrients and nano fertilizers can be used to enhance the growth and development of plants. We can supply nitrogen, P, MnO, CuO, iron oxide, SiO_2, and ZnO as nano-nutrition to the plants. The biological behavior and physico-chemical activity of the nanomaterials and nanoparticles must be studied in detail. These studies can be easier if models were used. The modeling process to study nanoparticles and nanomaterials will be a tedious and complicated process.

Several studies had been dealt with the nano fertilizer in cropping systems (Figure 8.2). The crop mineral nutrients can be in the form of nano fertilizers in two possible ways. The nanomaterials can be incorporated into a carrier. The elements like zeolite, clay, polyacrylic acid, and chitosan can be adsorbed or absorbed onto the carrier complex. The second method is to use the essential elements along with the carriers or just as in encapsulated forms or suspension. This second method can be used whilst the application of zinc, iron, and other elements onto the soils. These nano fertilizers have many advantages. They can offer rapid absorption rates and leaching will be minimized. The first method can be used as it offers comfort to the user. The nanoparticles can be used in the form of nano fertilizers. These substances can be readily absorbed by the roots. The absorption can be via symplastic and apoplastic pathways. These pathways can be absorbed in the xylem tissues. The nano fertilizers can be supplied to other parts of the plants through vascular bundles.

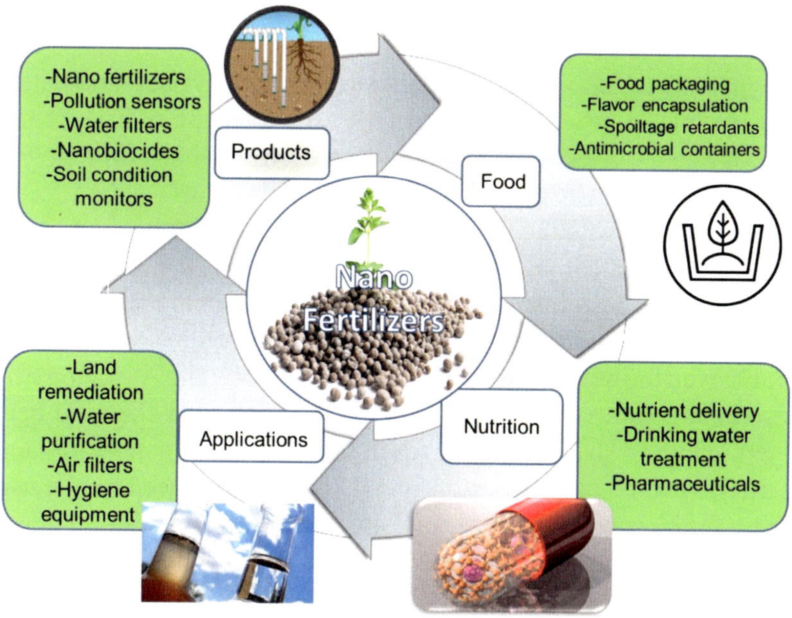

Figure 8.2: Nano fertilizers.

Source: https://link.springer.com/chapter/10.1007/978-3-319-96719-6_2

The fertilizer efficiency can be kept at minimal value. This can be applied when there are high N and P quantities. Most of the zeolites may not be classified as nanomaterials. Most of the effects of the nanoparticles will be based on size. Zeolites can be classified as safe for most soils and they can be used to enhance crop productivity. The anionic nutrients may not be maintained with the zeolites. These nutrients must be associated with the biopolymer complexes and biopolymers to compensate for this effect. The nanomaterials can enhance the absorption and transport of fertilizers. The fertilizer elements can be used in lesser quantities. The nano fertilizers can be absorbed by the plant roots easily. There are many classes of nanoparticles that can be transported to the cells via endocytosis. The channels and pores can be used by the plants to absorb the nanoparticles. The ZnO nanoparticles can be easily absorbed through roots. These nanoparticles may not pass the endodermis region. In the soils, nanofertilizers can permit improved dissolution. Due to this, there will be enhanced assimilation and absorption. We can add Mo, Cu, Zn, Mn, Fe, Mg, Ca, K, P, and N in the form of nano fertilizers. The nano fertilizers must be used in appropriate form and texture. There will be reduced competition, gasification, and leaching among the organisms.

The nanomaterials must get activated by a physical and chemical stimulus that plants provide. To control the excess release of nano fertilizers in the ecosystems, they should be associated with chitosan, biopolymers, films, and hydrogels. The ceramic materials and clay soils can be added with the fertilizers along with the nanocomposites. The organic nanoparticles can be encapsulated and can be used on crops. Emulsions can be used for this purpose. The seed germination rates were increased in the senescent seeds. Most of the seeds can get germinated even in unfavorable conditions. This can be achieved through the application of TiO_2, CuO, and Au nanoparticles. The carbon nanotubes can be used in this case. The seed coat will become more permeable so that water can reach the seeds. The nanomaterials and nanoparticles can be used in seed priming. These substances can be used in enhancing the stress tolerance rates in stressed crops.

Most of the nanoparticles of these elements may not be essential for the crops. The stress tolerance and the nutritional quality of the crops can be positively linked in mostcases. The crops that maintain a balance between the osmolytes, antioxidants, signaling and response, metabolites, and mineral nutrients can easily adapt to environmental stresses. This will induce specific characteristics in the plants so that they can get adapted to several negative stimuli. The tolerance of the plants that are suffering from environmental stresses can be enhanced if the aforesaid balance is maintained. There are many applications of nanoparticles and nanomaterials. Most of the plants face abiotic stresses that can be reduced by the application of the nano fertilizer and nanomaterials. In the case of conventional fertilizers, the elements that were classified as essential were added with the organic matter. The ground minerals that were used are dolomite and rock phosphate. The soluble mineral salts can be used in this case. The nano fertilizers can be transported from the crops to the soils within no time. They can reach the atmosphere and water bodies. They can be added to the runoff of the leachates. The transport wind can also carry these nano fertilizers. The application of nano fertilizers should be carefully monitored.

These substances should not damage human health, animals, and others. Most of the nanoparticles will be absorbed by the microbes that are present in the rhizosphere regions. These substances will be transported from the root system to the other parts of the plants. These particles may get accumulated and can reach higher trophic levels. This is a negative effect. This biomagnification effect was observed in marine organisms. The nanomaterials can get interact with viruses, magnetosomes, ferritins, lipoproteins, and exosomes. These substances mostly occur in living species. The nanoparticles can be produced by natural events like meteoric dust and volcanic activity. The crystallization, nucleation, and weathering process can produce many nanoparticles naturally. Most of the microbes can generate these nanoparticles naturally. It is to be noted that naturally occurring soils can harbor several nanoparticles in the form of iron oxides, organic matter, and clay. These nanoparticles are produced due to the natural events that occur near the crops. It is important to know that these nanoparticles can help in the growth and development of crops. The nanoparticles that are present in the soils can supply nutrients to the crop plants. There will be lowered leaching rates due to the presence of the nanoparticles. Some of the adverse reactions like decreased photosynthetic rate, enhanced oxidation stress, and chromosomal aberrations can occur in the plant systems. These adverse reactions in the plant systems can be controlled by genes.

The effect of the nanoparticles and their coating on the environment can be important in knowing the beneficial effects. The complexing agents and pH can be important in determining the effect of the nanoparticles in the soils. Environmental factors like light, wind, and temperature can influence the rate at which these nanoparticles interact with the crops and inflict their positive effects. The action of nanoparticles in the soils can be regulated by both biotic and abiotic factors. The toxicity of the nanomaterials can be direct and, in some cases, indirect. The toxicity of the nanoparticles can be prominent in the soil microbes. There will be an indirect effect due to the secretion of the root exudates. There will be a change in biodiversity due to the presence of nanoparticles. The adaptive capacity of the plants will be altered due to the modification brought by these substances. All these nanomaterials can bring about stress and environmental factors.

To determine the toxicity of these materials, there should be controlled conditions. The nanomaterials can induce toxicity in the crops in some cases and it is based on reactivity, diameter, and shape. These materials can enhance the performance of the crops. The nutritional quality of the crops, stress tolerance, and other factors can be affected by the nanomaterials. Most of the concepts that build the nanotechnology subject can be interdisciplinary. All the fields can participate in this area. The application of nanotechnology in the agricultural sectors has become popular. The nanodevices that can be used in the plant protection process can be called nanoparticles and nanocapsules (Figure 8.3). The nanocapsules can be enclosed with a shell. The shell can have agrochemicals that can save the plants from diseases and pests. These shells are made up of nano clays, viral capsids, lipids, and polymers. The shell protects the compound that is present inside. This can enhance the penetration and solubility and so the entry of the agrochemicals into the plant tissue can be easy. The compounds of the shell can determine the mobility of the active compounds present inside the shell. The shell will be sensitive to enzymatic changes and pH variations. This is based on the local environment and soil conditions.

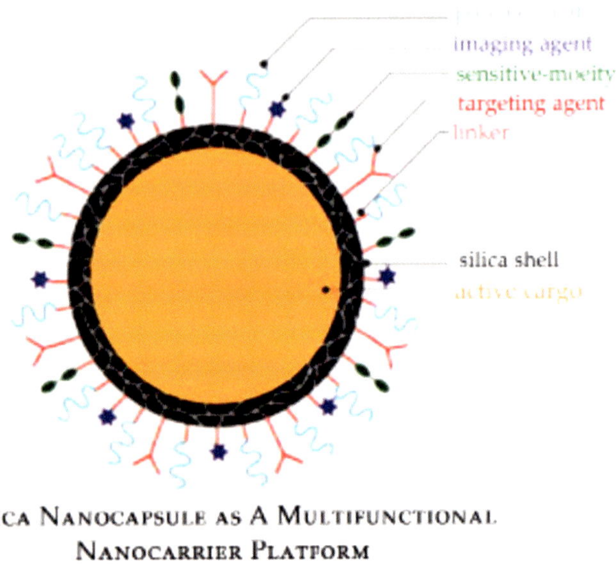

Figure 8.3: Nanocapsules.

Source: https://www.sciencedirect.com/science/article/abs/pii/S0001868616301117

The nanoparticles can have a solid core and, in some cases, we can use a solid core. This core can be prepared using polymers and metals. The nanoparticles will be encircled with biomolecules and linkers. Because of its minimal size, the surface area will be high. Due to this there will be enhanced biochemical reactivity. The superparamagnetism can be formed due to the physical properties of the nanoparticles. We can use nanodisks to supply several antimycotic chemicals. The nanodisk can be classified as a matrix. This matrix comprises phospholipids. This structure can shield the amphotericin molecules from pH and light. This structure can also enhance solubility rates. The active compounds that are present in the nanodisks can be sprayed directly on the leaves.

These nanoparticles and nanocapsules can have several advantages. Most of the active compounds can be protected by the capsules. These substances cannot get mixed with the soils generally. The nanoclays (Figure 8.4) can minimize the leaching process. So, the loss due to runoff can be reduced. The nanoparticles can be associated with the biomolecules so that the selectivity can be increased and so the targets will be appropriate. The nanomaterials can get accumulated in the animal tissues and vegetal parts. These nanoparticles can enter the food chains and can get accumulated in the trophic levels. Most of the nanodevices are expensive.

Figure 8.4: Nano clays.

Source: https://nanografi.com/blog/nano-clay-description-usage-and-applications/

The agriculture sectors need large amounts of nano products that will strain the economics of the growers. The plant genetic transformation can be based on nanotechnological techniques. The nanoparticles can be linked with the rDNA constructs. These constructs will be attached with the ligands so that they can enter the cell wall of the plant tissues. These nanoparticles can serve as vectors in the transformation experiments.

The genetic manipulation of the plants via nanoparticles can carry several advantages. There will be an increased transformation efficiency. The transformation efficiency will be higher than that of biolistics. The nanodevices can be used in both temporal and permanent genetic transformations. The gene silencing process can be done with the help of nanodevices. The genetic transformation can be easily done with these nanodevices. Agricultural nanotechnology can be used in organic agriculture practices. There will be sustainable production whilst using nanodevices. We can use nanocontainers that are made up of starch materials. These nanocontainers can be used in the delivery of nutrients to the surface of the plant leaves. We can use starch because it is classified as non-toxic, biodegradable, and biocompatible with the plant tissues and the environment. We can expect limited damage though. It is important to use starch that has a high amylose concentration. The amylose molecules can be linear and that determines the functional properties of the nanocontainers and starch that is being used. Starch can be used in minimized sizes. We can use amylose produced from the wheat varieties that can produce high amylose content. This can be done with the molecular mutagenesis processes.

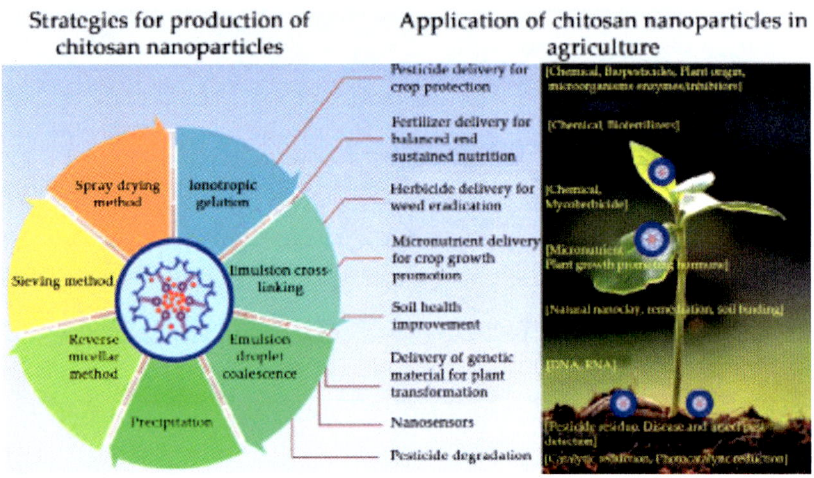

Figure 8.5: Nanoparticles.

Source: https://www.sciencedirect.com/science/article/abs/pii/S0141813015001257

The preparation and characterization of the starch nanoparticles can be important as it determines the effectiveness of the nanocontainers. We can use acid hydrolysis to generate nanoparticles from starch grains. This will demand more time. Due to this, there will be environmental stress and toxic wastes will be generated. We can use ultrasounds in this case so that the intervention of the chemical agents can be drastically reduced. The nanoparticles that are produced must be functional in every aspect. We need to modulate the nanoparticles enzymatically and chemically to get the desired results. With this step, we can expect the controlled action of these nanoparticles (Figure 8.5) on the crop plants.

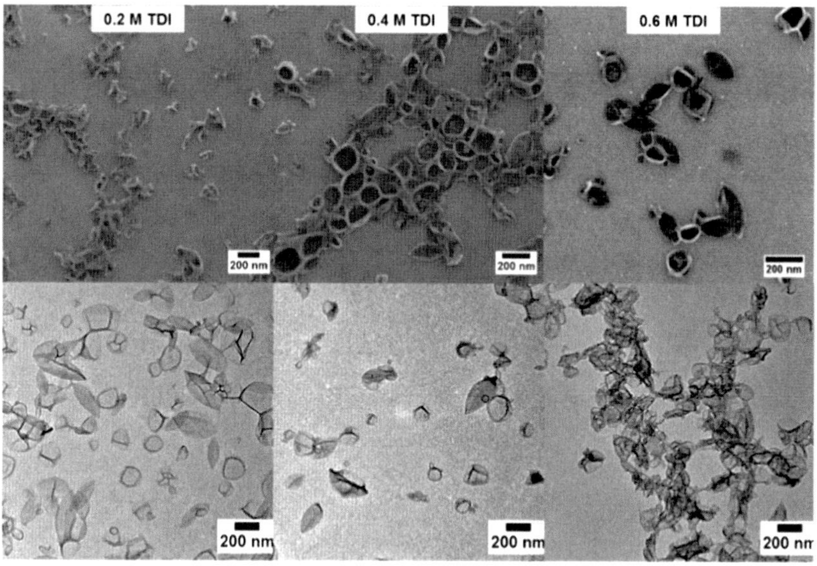

Figure 8.6: Nanocontainers.

Source: https://pubs.rsc.org/en/content/articlelanding/2014/ra/c3ra47971d#!divAbstract

Many kinds of molecules can be dispatched via starch nanocontainers (Figure 8.6). The nutrients can be supplied to the plants slowly so that they can absorb the essential nutrients for long periods. The essential micronutrients like zinc, manganese, and iron can be protected along with the phosphorous molecules. These nanoparticles can act as bio-stimulants and can be released onto the plant tissues when needed. This can be done with the enzymatic activity of the plant tissues. These nanoparticles must be protected from microbial attacks. Some of the advantages of these nanoparticles are minimal production costs, high quality, high selectivity, minimal use of chemical compounds, reduced harmful residues in the soils, and minimal phytotoxicity. Nanotechnology can be used in the production of nanodevices and production systems due to their unique properties pertaining to the magnetic, electric, mechanical, chemical, optical ad physical properties. Mostly these properties can be attributed to their size. We can use nanotechnology in the preparation of diagnostic devices. The quantum size effects are to be accounted for this unique property of the nanoparticles. The lower scale of the nanoparticles can permit these elements like zinc oxide, titanium oxide, carbon nanotubes and fullerenes to be used without any problem.

We can use carbon quantum dots, graphene, nanocavities, nanocolumns, nanoparticles, nanofibers, nanotubes, and fullerenes as nanostructures that can be used in the improvement of crop nutrient intake. We can use nanotechnology in the production of diagnostic devices and sensors for soil analysis. The bio-silica nanostructures can be used in manufacturing the sensors in the protein analysis. Biosilica can be produced by aquatic organisms mainly those live as marine organisms. The sponges and diatoms can be used in this case.

The nanostructured materials can be classified as fundamental components of nanoscience and nanotechnology, Nanoparticles can be called NPs. They have catalytic properties, optical, magnetic, and plasmonic properties. Most of the nanostructures exhibit high stability rates. This can be attributed to the high surface area. The physical multifunctionality of the NPs can be due to their lower scale. The nanostructured materials can be called as NMs. They can be used in the food industry, agriculture biosciences, and others. The nanofertilizers or NFs can be used in the nutrition of fruit crops. We can use these NFs in the form of sprays and soil-based formulations. The plants can be supplied with NFs loaded with the essential elements. The nutrients can be supplied to the plants with high efficiency. We can expect limited waste if we use NFs. Post entry into the cuticle and leaf tissues, the NFs can migrate to other plant tissues via hydrophilic, lipophilic, symplastic, and apoplastic interactions. All these interactions can enhance the effectiveness of the NFs. The NPs can interact with the plant cells and they can supply the nutrients via carrier proteins. They can bind to organic chemicals, endocytosis, ion channels and aquaporins.

NMs can exhibit high penetration capacity and diversity in shape. This can affect the growth and development of the plants. These NPs can enter the leaf tissues via stomata. The NMs can participate in several plant processes. The NPs can exhibit high penetration rates and this will hinder plant growth. The NPs must be used in minimal concentrations so that their negative effects can be reduced. The NMs must be encapsulated so that their concentrations can be safe and reach the plants in the appropriate time. We can use NFs in their encapsulated forms so that the formulations can be enhanced. There will be increased plant absorption. Nutrient loss will be limited. The encapsulation of the NFs can be done via three methods. We can use emulsions of NPs as delivery agents. We can coat the NFs with polymer films. Most of the NFs can be encapsulated within the NPs. The surface of the NMs can be modified. We can use nanocomposites in this case. The nanocomposites can have macronutrients like potassium, phosphorous, and nitrogen. The absorption of the micronutrients by the tomato plants is drastically increased on using the nanofertilizers. This will lower the loss of nutrients. The interaction of the nutrients with air, water, microbes, and soils can be reduced. The NMs can be applied directly onto the soils and plants in the case of strawberry crops. In some of the fruit trees, the NMs can be applied to the leaves and branches. Nanomaterials can be used in the packaging industries. Nanotechnology can enhance the shelf life of fruits. The health and freshness of the fruits can be enhanced. The coatings at the nanoscale can establish a carrier against enzymes, antioxidants, flavors, and colors from the external agencies. The nanopacking involves the use of chitosan and nano-silica. The nanopacking will ensure accurate preservation. The NFs can be used in horticulture crops. The NPs can exhibit high absorption rates.

8.2 BIOMASS AND YIELD

The carbon nanomaterials can exhibit thermal, mechanical, electrical, and chemical properties that are unique. The carbon nanomaterials can have single-walled carbon nanotubes and multi-walled nanotubes. The carbon nanotubes can encourage the elongation of the roots. This can be observed in the case of onion and cucumber crops. The length of the roots will be reduced in the tomato plants. Several studies considered metal-based nanoparticles. These nanoparticles are made up of molybdenum, iron, gold, copper, and silver molecules. These nanoparticles were applied on spinach, mustard, mung bean, maize, and wheat crops. The growth parameters of the plants were affected positively by the use of these nano fertilizers. These substances can reduce the abscission of the reproductive parts. The fertilizer use efficiency and water use efficiency will be increased. The formation of nodules will be increased. The seed germination process will be encouraged. The nanoparticles will induce positive effects on the protein contents, carbohydrate, chlorophyll, leaf area, and length of the roots and shoots.

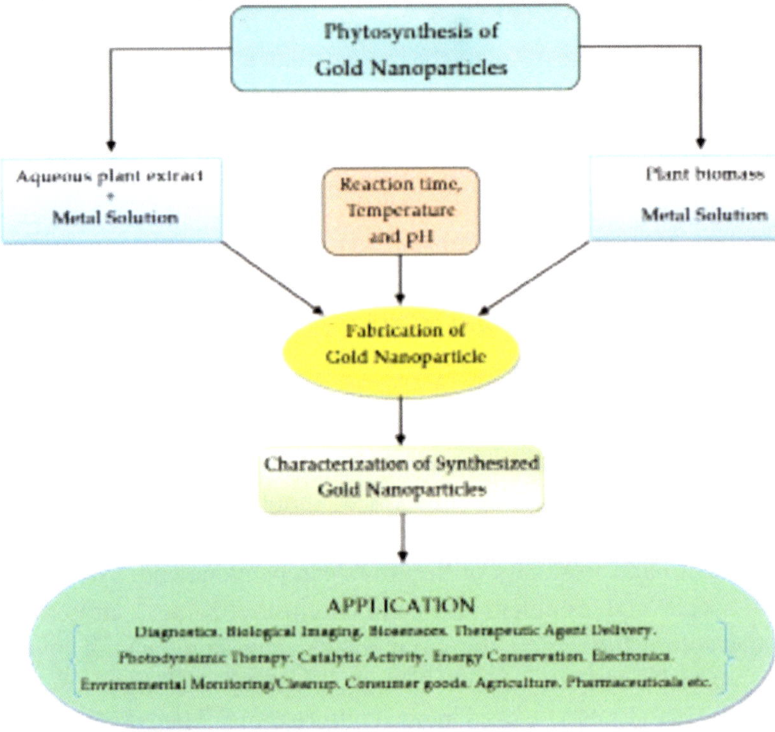

Figure 8.7: Gold nanoparticles.

Source: https://link.springer.com/chapter/10.1007/978-3-319-46835-8_17

The gold nanoparticles will be used in the genetic transformation studies in the plant cells. The gold nanoparticles can be used in crop improvement practices. The gold nanoparticles can be classified as essential tools to bring about a desired change in the crops. The application of metal-based nanoparticles of Mo, Cu, and Fe can have excellent application in the agriculture sector. Iron nanoparticles can be used in the growth and development of plants. These nanoparticles will affect the photosynthesis process (Figure 8.8).

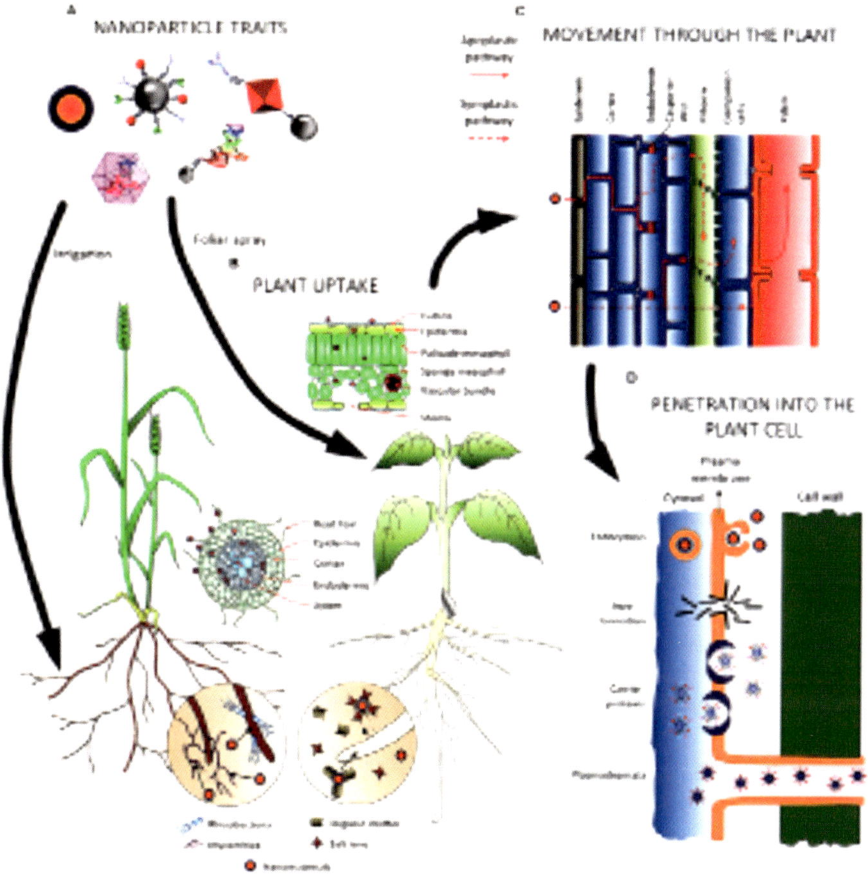

Figure 8.8: Penetration of nanoparticles.

Source: https://www.frontiersin.org/files/Articles/259755/fenvs-05-00012-HTML/image_m/fenvs-05-00012-g001.jpg

Most of the enzymes will get activated in the photosystems of photosynthesis machinery. It was observed that the biomass of the spinach enhanced on the application of iron nanoparticles. The accumulation of Fe in the roots, stem, and leaves has increased drastically. The ZnO nanoparticles can enhance the growth patterns in mung beans and chicken pea seedlings. The ZnO nanoparticles can be used to enhance seed productivity, flowering rates, and the growth of the plants. This can be observed in the onion crops also. In peanuts, the effect of the iron oxide nanoparticles was observed. These nanoparticles have enhanced the nutritional status of peanuts. In this crop, the rate of photosynthesis was also enhanced. In the case of soybean, there was an increase in the pod dry weight and leaf. The iron oxide nanoparticles can be applied via soil route or through foliar sprays. There will be enhanced photosynthetic rates and so the elongation of the roots. The absorption by the plants will be higher in the case of foliar sprays. Silicon dioxide nanoparticles can participate in the improvement of dry weight, fresh weight, vigor index, germination index, and seed germination rate.

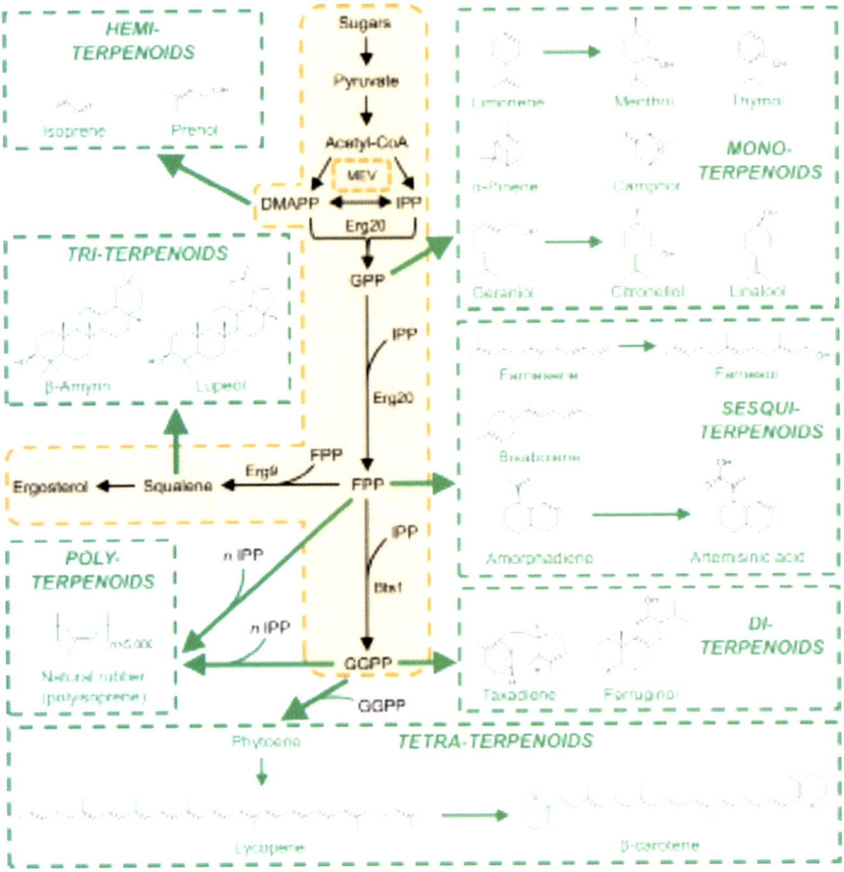

Figure 8.9: Secondary metabolites.

Source: http://www.plantphysiol.org/content/179/3/844

The secondary metabolite production (Figure 8.9) in the plants can be increased on the application of the nanoparticles. The phytochemicals will be affected due to the application of the nanoparticles. The secondary metabolites can be used in agrochemical formulations, pharmaceuticals, and nutraceuticals. The main problem with the secondary metabolites is that they are produced in limited quantities. We can use nanoparticles to drastically enhance the production of phytochemicals. The nanoparticles can be employed to transfer DNA fragments into the plant tissues. The silver NPs can exhibit high rates of mobility in soils that are dominantly negative charged. There will be long-term advantages with the silver NPs.

8.3 NANO FERTILIZERS

The plants can be improved due to the intervention of nanomaterials in the agriculture sectors. Mostly the improvement can be viewed genetically as the yields are increasing. The nano fertilizer technology can be classified as an innovative and useful technique. The nanoparticles can be employed as a regulator of the nutrients across the membranes. We can use nanocomposites and nano-fertilizers to regulate the movement of the nutrients. Due to this there will be enhanced nutrient use efficiency. These nanoparticles can avoid loss of nutrients and pollution of the soils. There are several attempts made to prepare the nitrogen nanomaterial formulations. The nanoclays can be derived from the montmorillonite and zeolite species. These species can release nitrogen molecules for extended periods than traditional fertilizers. The nanoparticles have enhanced the efficacy of these nutrients. The nano fertilizers can enhance fertilizer use efficiency in several crops. We can use nanoporous zeolites (Figure 8.10) that are available naturally in farming practices.

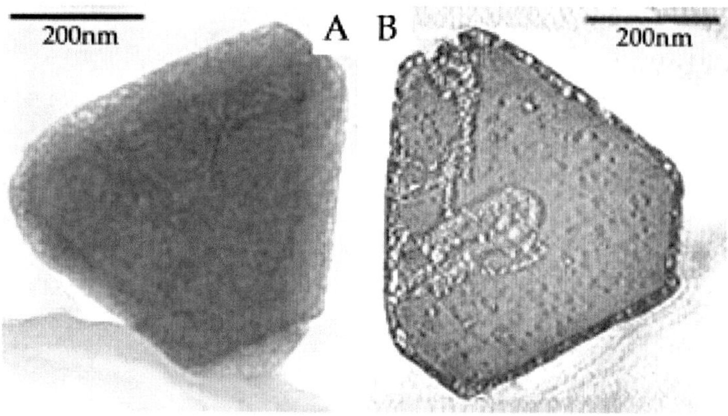

Figure 8.10: Nanoporous zeolites.

Source: https://link.springer.com/article/10.1007/s10562-014-1411-5

There will be minimal dangerous effects of these substances on the crops. Most of the agroecosystems will be less affected due to these nano fertilizers. Most of the crops can get benefitted from these nano fertilizers. The zeolites can occur naturally and there are about fifty mineral types that constitute the zeolites. It is always important to choose an appropriate growing medium that can support these nanoparticles. These nano fertilizers can exhibit a rigid 3D crystal configuration. There are channels and voids. There will be high CEC or cation exchange capacity. The clinoptilolite zeolites can comprise lattices that can harbor channels. These channels can adsorb water molecules due to their highly porous nature. The surface area is high and this property will help the nano fertilizer in maintaining their efficacy. The negative charge can help in holding the oppositely charged ions. There will be a magnetic attraction. Zeolites have CaO, K_2O, MgO, Na_2O, Fe_2O_3 and others that can provide essential nutrients to the plants. The nano-zeolites exhibit hexagonal symmetry. There will be 2D pores. The crystals can have cages and they have six-membered rings. These zeolites can support most of the positive and negative ions that are helpful for the plants.

The nitrogen molecules can be considered as important in the plant cell functioning. This is since all the metabolic, genetic and structural processes will be regulated by the nitrogen molecules either directly or indirectly. This molecule can be important in the chlorophyll functioning and structure. We can observe nitrogen in the amino acids so as in the proteins. Most of the proteins present in the plants can support the structure of the plants. ATP can comprise nitrogen compounds. Nitrogen molecules can be present in the nucleic acids that transmit the genetic information from one generation to the other generation. Most of the nitrogen molecules may not be absorbed by the plants. The nitrates that have a negative charge may not adhere to the soil particles with high affinity rates. To enhance the nitrogen molecules availability to the plants, sulfur coated urea, neem coated urea and polyolefin coated urea was used. The retention of these compounds in the soils and on the leaves can be important. This problem can be solved with the aid of zeolites that can enhance crop yields.

The phosphorus nanofertilizers (Figure 8.11) can be important in crop growth and development. P can be classified as an essential nutrient as it can support the plant skeleton. These P molecules can aid in several biochemical reactions that occur in the plant systems. P can help in energy transfer reactions. The phosphates like ATP and ADP are important molecules as far as the energy requirements are concerned. Most of the metabolic substrates, nucleic acids, coenzymes, sugar phosphates, phospholipids and proteins can have phosphorous molecules. It is important to supply P to the crops in their initial growth stages. The reproductive system of the plants will be affected by the P molecules. On the application of P molecules, the plants will attain enhanced resistance towards diseases, crop quality will be enhanced and the nitrogen-fixing capacity will be improved. The phosphate rocks and zeolites can exhibit slow release fertilization in the soils. This property can be due to dissolution reactions of the ion exchange. The zeolites can extract the calcium ions from the PR or the phosphate rocks. There will be a production of ammonium and phosphate ions. The nano-formulations of the phosphates can be applied to the crops so that the phosphates can be available to the crops for extended periods.

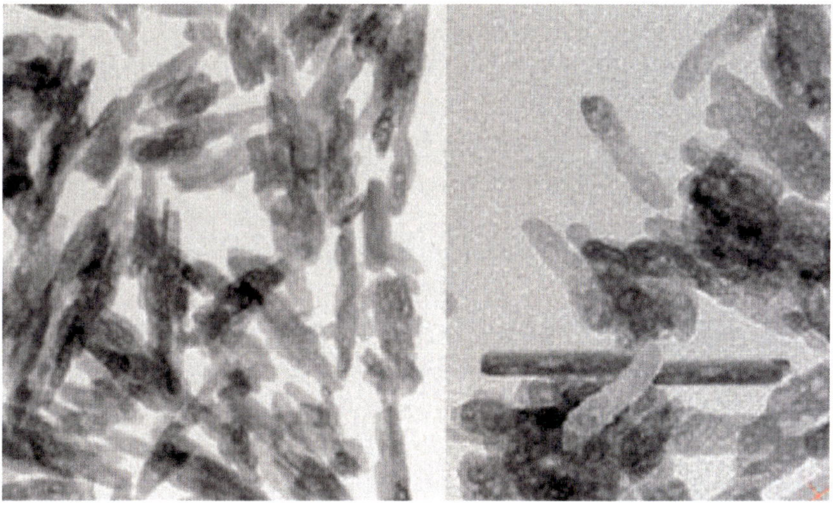

Figure 8.11: Phosphorus nano fertilizers.

Source: https://scialert.net/fulltext/?doi=jps.2016.52.60

The release rate of the P can be based on the P composition of the zeolites. P can be released from the rock if the pH values are lowered. Most of the ammonium ions can be converted into nitrates if pH values are low. We can use zeoponics if the plant needs N, P, and K. This process combines ion exchange reactions and dissolution. Post absorption by the roots, the nutrients will be rejuvenated in the zeolites. Most of the zeolites will be recharged with the supply of nutrients. Most of the zeoponic systems can enhance the retention of nutrients. Due to this, the loss of nutrients will be reduced. The need for excess fertilizer use will be reduced. There will be a balanced nutrient supply in the rhizosphere zone.

Potassium nano fertilizers have helped many growers in attaining better crop yields. Potassium can be considered an important factor in many plant processes. It may not be considered an essential element in the plant structures. It mostly participates in the regulatory reactions. lant growth and development can be greatly affected by potassium ions. K can participate in enzymatic reactions, stomatal regulation, ionic balance, protein synthesis and photosynthesis. K can affect six important enzymes that are important in the growth of the plants. The K deficiency can leave the plant unprotected from several pests and diseases. The plants suffer from drought stress. The K deficiency in the plants can render them susceptible to extreme temperature variations. The plants that are K deficient can be susceptible to nematode attacks. The K ions can affect the fiber quality, shelf life, taste, color, shape and size of the plants. The zeolites can produce enough K^+ ions that are needed for plant growth in lab conditions. The K zeolites can release the K ions slowly into the soils. These substances can adsorb most of the nutrients that are being supplied from the external aids. Most of the natural zeolites can be selective for potassium ions. They may not attract magnesium and calcium ions because they are divalent. The zeolites can be classified as important nutrients for plants as they attract anions and cations.

Magnesium, calcium, and sulfur can be classified as secondary nutrients. These nutrients can be needed by the plants in high quantities. These nutrients can be used in the growth and development of plants. The SMZ or surfactant-modified zeolites were used as additives in the fertilizer. These SMZs can be classified as essentially good carriers for sulfate ions. We can expect a slow release of sulfates if SMZs are used. Most of the nutrients that are required in lesser amounts can be called micronutrients. These micronutrients can be important in the growth and development of plants. The micronutrient deficiency in the plants is common in most Asian countries. This is since in these countries imbalanced NPK fertilizers were used. They also use high bicarbonate content and there will be salt stress in these regions. The bentonite and chabazite are classified under natural zeolites. These substances can absorb iron and zinc molecules. These zeolites can allow the slow release of the fertilizers. The cationic micronutrients are more due to the presence of the zeolites.

Nanocomposites can play important role in agricultural nanotechnology. We can produce new materials that can exhibit new properties. The P and N ions can be released simultaneously into the soils if we use phosphate rock and clinoptilolite. The nanocomposite (Figure 8.12) films can comprise nanolayers that are inorganic. These nanolayers are made up of silicate layers.

They are organic polymers and clay polymers. Zeolite can be classified as an effective carrier. It can be classified as a regulator and stabilizer of fertilizers. These nanocomposites can allow us to use limited amounts of fertilizer in the crops. On the application of the nanocomposites, there was an improvement in the grain yield of wheat. The protein concentration was also increased. This effect is better than NPK chemical fertilizers. The nanocomposite films can comprise inorganic nanolayers. Nanotechnology can regulate crop improvement mechanisms. We can use nanoparticulate formulations to enhance crop growth. The zeolites can be linked with fertilizers so that the nutrients can be retained. These nano- particulates can enhance the soil quality for long periods. On the application of these particles, there will be enhanced absorption.

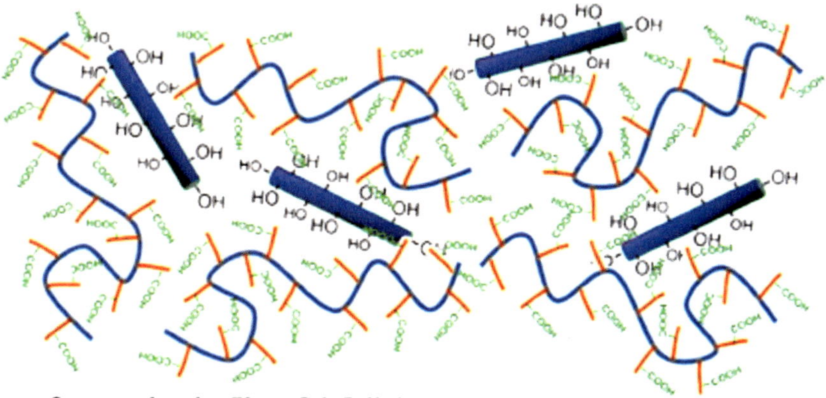

Supramolecular Plant Oil-Cellulose Nanocrystal Nanocomposites

Figure 8.12: Nanocomposites.

Source: http://www.polymer.cn/ss/wangzhongkai/en/publicationsshow_12253.html

The zeolites can hold the nutrients and retain them for long periods. Every year, there are huge losses with the excessive use of fertilizers and this can be attributed to the leaching process that may involve the rhizosphere zones. Mostly this can occur in sandy soils. The sandy soils may not retain the nutrients for extended periods. These zeolites can be associated with chemical fertilizers and there will be better results. This improvement was seen in grape and peach crops. The nano zeolites can provide appropriate amounts of calcium to the soil-plant continuum. Due to this, there will be a lowered sodium and calcium ratio. Their calcium ions will be derived from the zeolites in the rhizosphere media. Due to this, there will be excess deposition of the sodium ions. There will be reduced-sodium concentrations in the shoots. The sodium-calcium ratios will be altered in the roots and shoots. The carbon nanotubes can be added to the seeds so that the seeds can germinate at accelerated rates. The weight of the seeds will be increased or improved by using nanomaterials. The carbon nanotubes can allow effective penetration rates in the seeds.

8.4 NANOPESTICIDES

The applications of nanotechnology in plant nutrition and protection have become popular in recent years. These nanotechnologies-based products were classified as nanoagrochemicals. Agrochemicals can be considered important in modern agricultural practices. These nanoagrochemicals can be of great concern because of the environmental impacts. Agrochemicals are being used in agriculture with immense benefits for decades at expense of human health.

These chemicals can contaminate food and water resources. The application of nanotools not only reduces the pollution rates but also regulates the inputs drastically. Precision agriculture is possible with nanotools and associated materials. Most of the inputs pertaining to water, fertilizers, and pesticides will be regulated by the intervention of se nano-based tools and methods. These devices can enhance the quality of agricultural practices. The quality and quantity of the crop yields can be enhanced with the intervention of nanotechnology-based techniques. It is sensed that agriculture is a low-profit practice that seldom gets yields financially. Most of the crops grown with the application of excess chemical formulations faced social rejections and regulatory penalties. The Nanopesticides can replace the gaps produced by the insecticides. These are classified as chemicals that can protect plants. They can act on the crops via carrier molecules and active ingredients. Most of the environmental hazards of agriculture can be minimized.

This can be attributed to the small size of the Nanopesticides. The surface area of these nano pesticides is much higher so that most of the pests can ingest these materials. There will be surfactants, carrier molecules, and active ingredients in these nano pesticides structures. The mobility of these particles will be high. The mode of action will be enhanced and that is not possible with the conventional methods of pesticide utilization. Most of the nano-formulations can have special properties. There will be enhanced soil properties and the foliar spray can fetch excellent results. There are several variants in nanotechnology-based products.

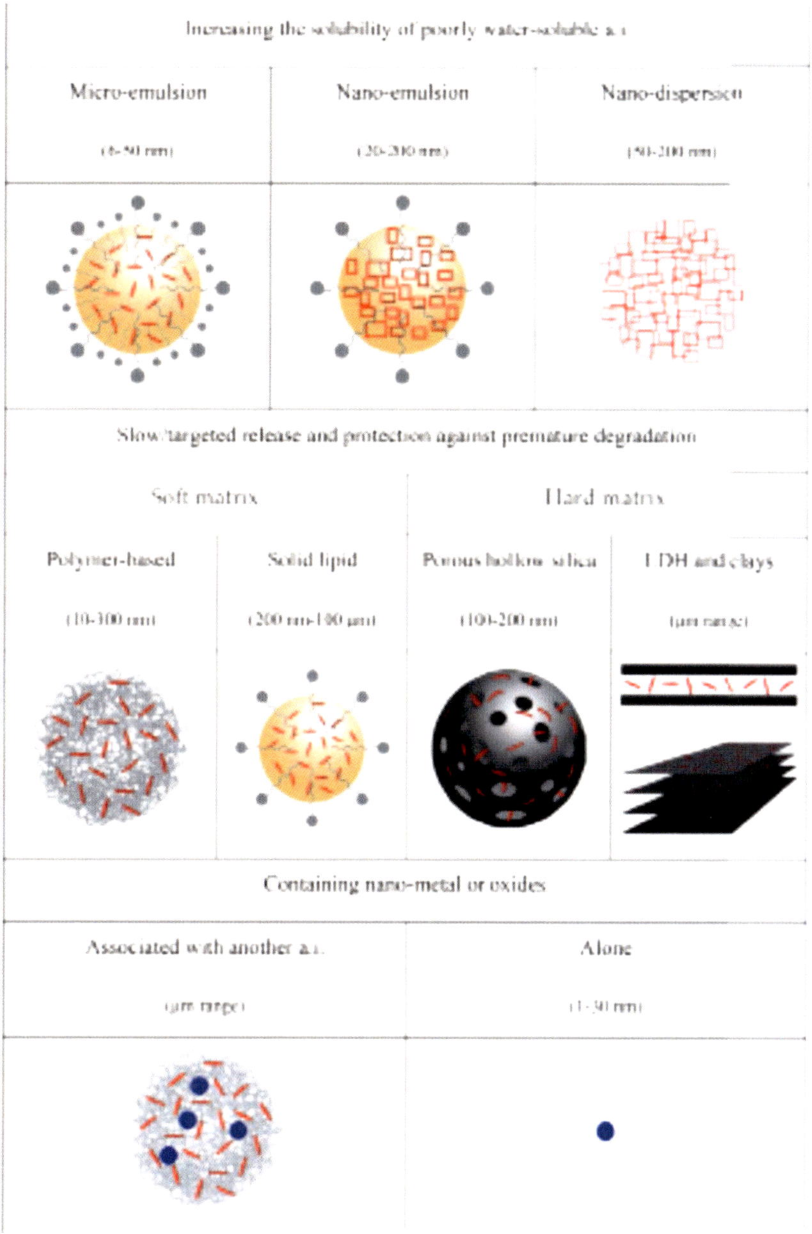

Figure 8.13: Nanopesticides.

Source: https://www.researchgate.net/publication/254216791_Nanopesticides_State_of_Knowledge_Environmental_Fate_and_Exposure_Modeling/figures?lo=1

They are nano-nematicide, nano-fungicides, nano-herbicide, and nano-insecticides. There are some advantages to nano pesticides. They cannot get degraded easily. The release of the active ingredients will be slow. There will be enhanced solubility. Some of the formulations with which the nanomaterial-based techniques can be used are clay-based nanomaterials, metal oxides, lipid-based formulations, and organic polymer-based formulations. The nano-emulsions can be considered as important materials that need to be applied whilst experimenting with crop nanotechnology. The chemical ingredients immersed in the nanosized water droplets are called nano-emulsions. The nano-emulsions can be classified based on the surfactants and the quantity they carry. They can be kinetically stable and thermodynamically stable. The pesticides can get soluble in water. This can be possible only if the water molecules, pesticides, and surfactants get fused. Then we call these materials thermodynamically stable emulsions or nano-emulsions with thermal stability. Tween 20 is being used as a surfactant in nano-emulsions. These nano-emulsions can control pests.

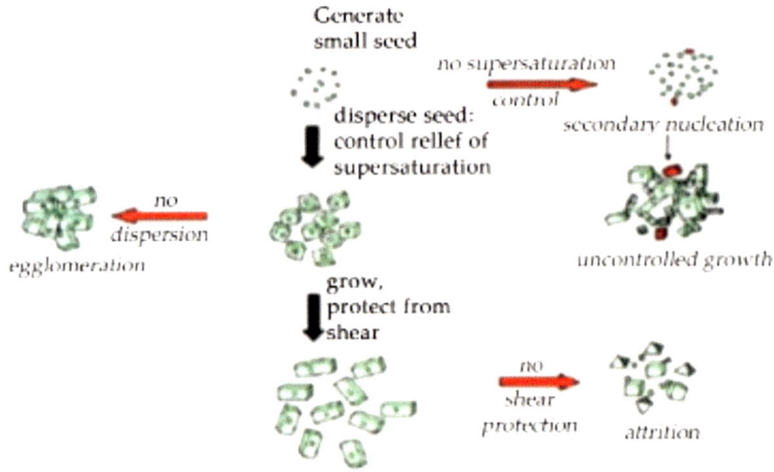

Figure 8.14: Nanosuspensions.

Source: https://scientiaricerca.com/srcops/SRCOPS-01-00020.php

The nanosuspensions (Figure 8.14) also called as nano-dispersions. These materials can disperse the pesticides and they will be solid particles at the nanoscale and are present in aqueous media. The surfactant molecules can be present on the surface of the particles. The polar side of these molecules will be into the aqueous solutions. The non-polar side can be linked with the solid portion of the pesticide. Beta-cypermethrin linked with novaluron can be classified as nano-suspension. We can use polymer bases nanoparticles as carriers of the pesticides. There will be a slow release of the ingredients so that the pesticides can reach the target site within stipulated times. There will be enhanced dispersion in the media that is aqueous. We can classify it as a protective reservoir. Some of the nano-based entities that can be classified under this category are nanofibers, nano-gels, nanospheres, and nano encapsulation. The nanocapsules can be classified as a heterogenous type of reservoir. The active agents can be hydrophilic or hydrophobic and can be restricted to the inner core cavity. There will be a polymer coating on these encapsulations. Azadiractin can be extracted from the neem plants. This formulation can be shielded effectively when used in the nano-scaled dimensions. Neonicotinoids can be classified as an important nano-formulation in crops.

Nano-farming (Figure 8.15) is important in the development of agriculture. The nanoparticles can enhance the characteristics of several plants in a positive way. Most of the nanoparticles can be classified as three-dimensional objects. In this particle, two dimensions can be in nanoscale and one dimension can be in macroscale.

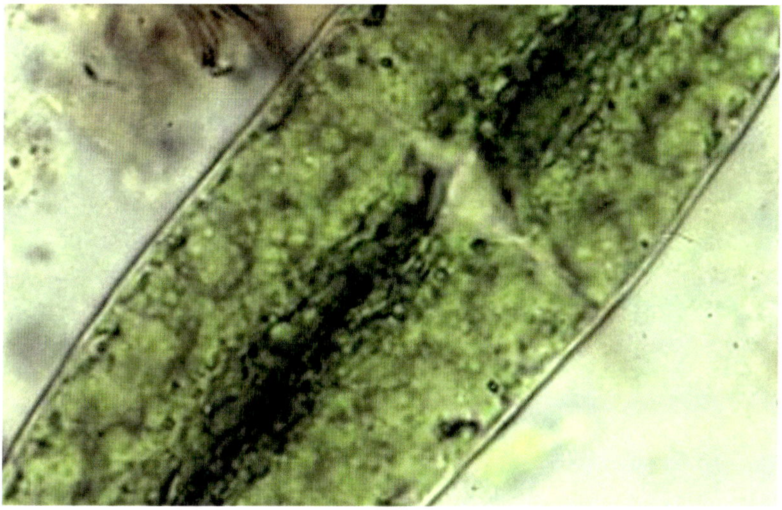

Figure 8.15: Nanofarming.

Source: https://ecofriend.com/eco-tech-nanofarming-produces-biofuel-from-algae-without-killing-it.html

The macroscale entities can include nanotubes and nanowires. The germination of the seeds can be greatly affected by soil fertility, moisture, genetic traits, and other environmental factors. The multiwalled carbon nanotubes or MWCNTs can affect the germination of the seeds. The germination of the seeds due to the application of the nanoparticles can be observed in garlic, peanut, maize, barley, soybean, corn, and tomato (Khodakovskaya et al., 2009 and Joshi et al., 2018). The zeolites, titanium dioxide, and silicon dioxide can stimulate seed germination activities. The silicon dioxide nanomaterials can act on the seed germination stages of the maize and barley crops (Disfani et al., 2016). The nanomaterials can penetrate the seed coats and so that there will be an effective utilization of water. The nanomaterials can stimulate several enzymes in the seed germination stages (Neetu et al., 2018 and Kumar et al., 2016). The hydroxyfullerenes (Figure 8.16) can enhance the growth and development of the crops. These nanoparticles can be used on mustard, potato, tomato, spinach, onion, wheat, etc. The fullerenes that have OH groups can cause positive effects on the growth and development mechanisms of the plants (Gao et al., 2011).

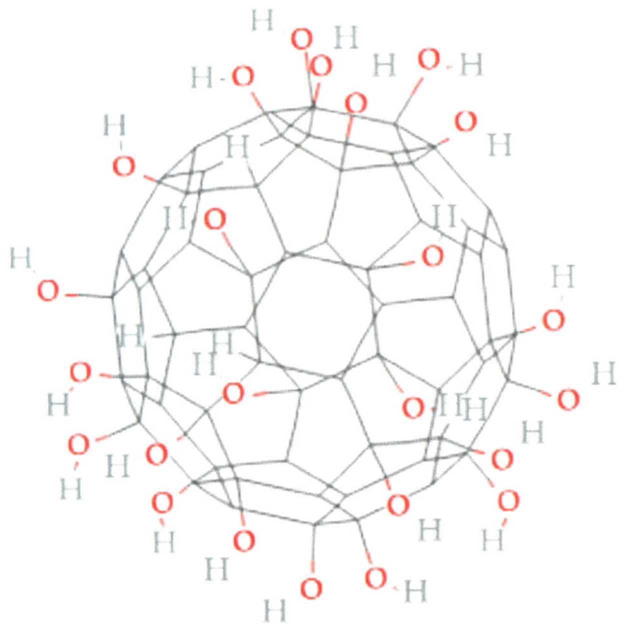

Figure 8.16: Hydroxyfullerenes.

Source: https://ecofriend.com/eco-tech-nanofarming-produces-biofuel-from-algae-without-killing-it.html

In *Arabidopsis* species, the fullerenes have increased the growth of the hypocotyls and this was possible with the stimulated cell divisions. We can use fullernol to enhance the size and number of fruits (Kole et al., 2013). Boron fertilizers and nano-zinc can be used to increase fruit yields (Davarpanah et al., 2016). There will be increased enzymatic activity on the application of nanomaterials (Dubey et al., 2016 and Lyons et al., 2010). The mode of application, surface coating, shape, and size of the nanomaterials can be important in understanding the role of these materials in the growth and development of the crops (Cox et al., 2016). It was observed that the performance of the nanomaterials is based on the mode of the nanomaterial application and use. The foliar type of application can enhance the dry weight, fresh weight, branches, leaves, the height of the plant, oils, total carbohydrate content, and total chlorophyll content.

Food security has been a burning issue for decades. Climate change has decreased most of the food production yields. Through nanotechnology, we can enhance the adaptation mechanisms of the crops easily. Advances in nanotechnology have resulted in an enhanced rate of adaptation (Onaga et al., 2016). The silicon nanoparticles can reduce the stress caused due to UV-B radiation in wheat crops (Tripathi et al., 2020). The nano-zeolites can enhance seed germination rates (Manjaiah et al., 2019). The nanomaterials can be used in the remediation and detoxification of the pollutants that can affect crops and humans. The silicon nanoparticles when applied to the rice plants reduced cadmium stress in them. The nano silicon fertilizers can have several advantages over conventional fertilizers in regulating the accumulation of heavy metals (Wang et al., 2014). In several cases, the productivity of the crops will depend on the biotic factors (Cox et al., 2017). Nanocomposites can be used to protect plants and they are classified as ecofriendly. The chitosan nanocomposites can be associated with the silver nanoparticles and there was enhanced antifungal activity (Le et al., 2019). The increased enzymatic activity due to the application of the nanomaterials was studied in detail (Shojaei et al., 2019). The nano zinc oxide and nano silicon dioxide can enhance the deposition of the amino acids and proline. The activity of the glutathione reductase, nitrate reductase, peroxidase, catalase, and superoxide dismutase was altered on the application of these compounds. The nanomaterials can also regulate the genes that are needed for stress resistance and tolerance. The silver nanoparticles can down-regulate or up-regulate the genes in *Arabidopsis* . The nano-materials directly participate in inducing stress resistance in several plant species. The nano fertilizers can affect the plants in different stages of their life cycle (Abdel et al., 2016). Generally, plants respond to the stress conditions via physiological variations induced by hormones like salicylic acid, methyl jasmonate, and jasmonic acid. The nano-formulations of pesticides have several advantages over traditional pesticides. The coating nanomaterials play a crucial role in inflicting damage to the pests of several crops (Nuruzzaman et al., 2016). The nanosphere formulation of silica can enhance the tendency of the pesticides to enter plants and then cell sap. Due to this, there will be a systemic effect to regulate aphids. Certain hollow formulations can protect the pesticides against sun radiation (Panpatte et al., 2016). The nanoformulations can change pesticide behavior (Hou et al., 2016). The nano copper materials can damage Phytophthora species that can damage tomato. The nano-scale herbicides can stop weed growth.

8.5 APPLICATIONS OF NANOTECHNOLOGY

It is important to understand food security and sustainable crop production in the field of agriculture (Harikesh et al., 2016, Kou et al., 2017). In the last two decades, there was a significant advancement in the field of nanotechnology especially in the agriculture sector (Man et al., 2018, You et al., 2016). ustainable agriculture demanded agrochemicals and it is important to attempt nanotechnology to dilute this situation. The primary problem with the plants is their adaptability against the perils of climate change (Vermeulen et al., 2012). limate change has already dwindled the equilibrium of the sensitive ecosystems.

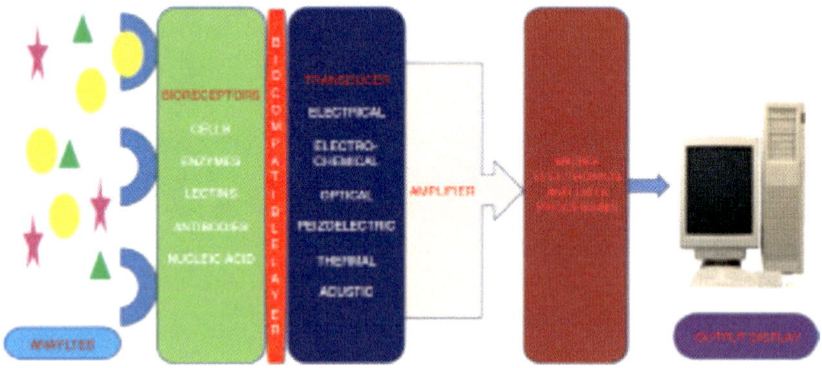

Figure 8.17: Nanosensors.

Source: https://link.springer.com/chapter/10.1007/978-981-10-4573-8_13

Nanosensors (Figure 8.17) are used in agricultural practices to ensure better yields by regulating the progression of a disease and pathogen. Nano-engineering has become an important tool to enhance crop production in the case of many agroecosystems. Nanotechnology not only coordinates agriculture sectors but manages well than conventional technologies. Nanotechnology-this term was coined by Norio Taniguchi in 1974 (Khan et al., 2014). This technology is still in its progressive stages and a lot of research is due (Bucheli et al., 2012). CRISPR or clustered regularly interspaced short palindromic repeats mRNA has become an important achievement in nanotechnology-based genetic research (Ran et al., 2017). The nanoproducts can answer many questions posed by environmental extremes. They can protect plants from diseases and environmental stresses (Conor et al., 2016).

The nanoparticles or NPs can be classified as hybrid materials, inorganic and organic materials, and that one dimension are in nanoscale. Most of these NPs can occur naturally. They can be produced from microbes, animals, plants, erosion, forest fires, volcanoes, and photochemical reactions (Eva et al., 2016). These nanoparticles can be ecofriendly and we can define them as least risk products (Marsili et al., 2014). The zero-dimensional nanoparticles can be defined as materials with three dimensions at the nanoscale (Kim et al., 2012). The quantum dots can be important in this case as they are classified as zero-dimensional nanoparticles (Wonjoo et al., 2009).

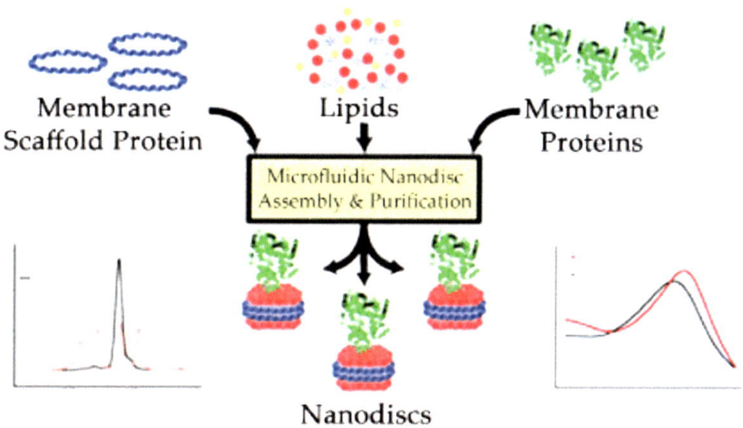

Figure 8.18: Nanodisks.

Source: https://pubs.rsc.org/en/content/articlelanding/2017/lc/c7lc00601b#!divAbstract

Nanodisks, nanowalls, nanosheets and nanoplates can be important (Figure 8.18). Some of the advanced products of the nanoparticles are nanoreactors, nanocontainers, photocatalysts and sensors (Leung et al., 2010). With this technology, it is technically possible to develop sustainable environmental systems. There will be reduced inputs in agriculture with the intervention of nanomaterials and so the wastes will be reduced or in some cases eliminated (Josef et al., 2015). The delivery mechanism of DNA via nanoparticles can be important as it is important for sustainable agricultural development (Olivier et al., 2016). Most of the losses incurred in agriculture can be due to microbial degradation, hydrolysis, photolysis, and leaching of the chemicals (Xinmin et al., 2015). The delivery techniques and equipment can be reduced whilst the application of nanoagrochemicals. In this case, a nano-matrix enclosed with active ingredients can be used for effective delivery. The cost of producing nano-based agro compounds can be less. This will help the growers at large. The nanomaterials can enhance the stability of the active ingredients and protect them from environmental factors.

The nanomaterials can be used to later the genes and so we can easily produce hybrids. In the case of nanobiotechnology, we can use nanofibers, nanocapsules, and nanoparticles to transfer DNA to the species of interest for better yields. Due to the advancements in nanotechnology, genetic materials can be easily transferred now without any problem (Torney et al., 2007). We can use nanotechnology in the genetic engineering of corn and tobacco plants. The nanoparticles can be coated with the nanoparticles and this is called as gene-gun technology (Periyasamy et al., 2010). Chitosan can bind with RNA and they can be linked with the nanoparticles so that the entry into the cell membranes will be easier. Nuclear gene editing has become much effective in maintaining the cell lines in the production of the crops (Rubul et al., 2017). It is very important to use element fertilization so that the fertility of the soils and crop productivity will be enhanced. Fertilizer management can be classified as an important step in managing sustainable agriculture. To feed the growing population, there should be at least a 70% increase in crop production. Most of the agricultural lands are either damaged or on the verge of becoming barren lands.

8.6 MANAGEMENT OF PESTS

The regulation of vectors, insect pests, and parasites needs eco-friendly tools that need to be more effective. There will be resistance developed against pesticides and drugs. The novel bio-active compounds can inflict diverse modes of actions. The risk to the environment and human health must be less. There are many natural products that can regulate vectors, insect pests and parasites. This can be due to their high rate of efficacy. Most of these natural products may not be target-specific but they act on almost all the pests. We can add these natural products to the nanosystems and so their physicochemical properties will be available for us to try new methods of control. The nanoemulsions and microemulsions can be prepared industrially on a large scale. Most of the preparation strategies can be simplified. The nanoemulsions and microemulsions can be classified as important tools in the management of vectors and pests. The chronic toxicity of the species that are not previously classified as targets makes them effective.

The agricultural sector, food, and pharmaceutical sectors needed delivery systems that can take the compounds onto the target cells without environmental stress or damage. These tools can be in the form of colloidal dispersions. In this case, the inner phase can be dispersed into the non-discrete medium. Most of the nanoemulsions and microemulsions can be classified as auto-emulsifying systems. The inner phase can be about 100 nanometers (Anton et al., 2011). Due to this, there will be enhanced bioavailability and stability. Some of the phenomena like sedimentation and creaming can be avoided by these nanomaterials. The effect of gravity on the nanomaterials is quite less. There will be minimal interfacial tensions and surface tensions so that the penetration of the bio-active compounds will be easier (Tadros et al., 2004). Both nanoemulsions and microemulsions can comprise surfactant agents, cosurfactant, oil phase and aqueous phase. This is the main reason behind their easy incorporation into lipophilic and hydrophilic compounds (Talegaonkar et al., 2008).

The components that can be used must be carefully planned. We can use vegetable and mineral oils as oil phases. In this case, soybean and sunflower oil can be used. The oil phase can carry lipophilic molecules. The basic aspect of the essential oil formulations can be based on the choosing criteria of surfactant agents. The surfactants have amphiphilic values and mostly we can derive those properties in terms of lipophilic and hydrophilic balance. This balance can be based on the continuous phase. We can prefer intermediate values. When intermediate values are chosen, we can get partition between the two phases and so most of the interfacial tensions will be reduced. With this, there will be droplet stabilization and their formation. These surfactants can be classified into the following:

- Anionic surfactant
- Cationic surfactant
- Non-ionic surfactant
- Zwitterionic surfactant

These substances have several applications and they can be produced on a large scale as the formulation is quite easier. The physicochemical properties and applications can be important. The microemulsions should have the surfactant to oil ratio greater than 2 whereas this can vary between 1 and 2 in the case of nanoemulsions. The particle dimension can be in the nanometric range. The free energy of the microemulsions can be less than 0 and in the case of nanoemulsions, the free energy will be greater than zero. The microemulsions can be classified as thermodynamically stable entities and the nanoemulsions can be kinetically stable. The fabrication of the microemulsions can be due to spontaneous formation. The fabrication of nanoemulsions can be possible with high-energy methods like mechanical devices. The long-term stability of the microemulsions can be due to their indefinite kinetic stability.

In the case of nanoemulsions, the height of the energy barriers will be important. The barriers will be between the separated states and nanoemulsions. The microemulsions can be thermodynamically stable liquid with optical isotropic nature (Lawrence et al., 1994). The microemulsions were first observed to recover oils from the reservoirs that were underground (Hoar et al., 1943 and Bera et al., 2014). The use of microemulsions spread to many fields. They have diverse applications in pharmaceuticals. Since then, these emulsions were used in drug delivery systems (Pavoni et al., 2019). The surfactants when used in high proportions can be toxic to the environment (Venhuis et al., 2004 and Wilhelm et al., 1993). The nanoemulsions can be classified as unstable colloidal dispersion thermodynamically. In this case, there will be two immiscible liquids. The free energy present in these compounds can be considered an important property. Due to this, there processes like formulation, preparation and even stability. Most of the microemulsions can be thermodynamically stable. The free energy can be due to the separate state of oil and water in the colloidal systems. Most of the microemulsions can be favored energetically. The nanoemulsions can exhibit a high rate of free energy. The microemulsions can be produced by the mixing of surfactant, water and oil randomly. There is no need for energy to be supplied from the outside. We can use heating or magnetic stirring in this case. The free energy can be negative so that the entire process may occur spontaneously. Due to low interfacial tensions, there will be an interaction between the oil phase and hydrophobic tails. Due to this, there will be the production of an oil-water interface (Schulman et al., 1959). In some cases, we may need to add cosurfactants like alcohols. This can reduce the formation of excess surfactants (Lam et al., 1987). Most of the microemulsions can be classified as dynamic systems. The interface can be in motion according to the Brownian motion and so there will be alterations in their respective radii. In the case of nanoemulsions, the free energy will be positive and they are thermodynamically stable. There is a need for external energy.

Based on the physicochemical interactions and possible mechanisms, the preparation of the nanoemulsions can be classified into low energy and high energy methods. The high-energy mode of preparation can be based on mechanical devices. The mechanical force can produce the droplets via dispersion. The droplets have a diameter lesser than 200nm. We need to produce macroemulsions with the aid of a mechanical stirrer. The macroemulsions can be transformed into nanoemulsions. In this process, we can use a homogenizer, sonicator, and microfluidizer. These devices can produce minute droplets from the larger ones. We can use ultrasound waves for the same effect. We can use the cavitation process for that matter.

There are certain limitations pertaining to costs and industrial processes (Silvia et al., 2019). The methods that may need low energy can be simpler and cheaper. There will be the production of minute droplets. To achieve this, we need to have enough knowledge of all the parameters involved (Chang et al., 2014). Free energy can affect the stability of the materials for long periods. The microemulsions can be stable. The nanoemulsions can be in the metastable phase. We need to understand the instability phenomenon like flocculation and coalescence. In the case of coalescence, the small droplets will be fused to form larger ones. This can be linked with the surfactant layers that mostly form on the droplets. The nanoemulsions are devoid of flocculation and coalescence. Most of the nanoemulsions can be susceptible to Ostwald ripening. This is nothing but the disappearance of the minute droplets as they merge. Most of the small particles can be soluble and this can be attributed to their larger radius of curvature. The inhibitors of the ripening process are sunflower oil, corn oils, and medium-chain triglycerides (Terjung et al., 2012).

The stability of the nanomaterial can be based on the relative concentrations. These concentrations can affect the interfacial tensions. It is impossible to maintain the stability of the parameters. The stability of the parameters can be based on the native characteristics of the compounds. The stability of these substances can be based on the ionic strength of the medium. It determines the repulsive forces between the phase of dispersion and droplets. Once there is an increase in the ionic strength, there will be a decrease in the repulsive forces. Due to this, the system will remain unstable. The rate of solubility in the dispersed phase can be considered an important factor in determining the stability of the nanomaterials. Due to solubility, the droplets can migrate to the continuous phase. There will be the presence of Ostwald ripening. Temperature can be considered an important factor in determining the stability of these substances. The solubility rates can be affected. The energy balances can be affected in the system. The nanosystems can be used in many chemical applications, fuel, and oil recovery operations (Shah et al., 1998). The nanoemulsions can be classified as a basic tool in delivering functional substances in beverages and functional foods (Vittal et al., 2019). The nanoemulsions and microemulsions can be used in encapsulating nutrients so that they may not get damaged due to environmental stresses like light and temperature. These substances can enhance the shelf life of the foods (Donsi et al., 2012).

The nanoemulsions and microemulsions can be present in the food chains but with limitations intact. This can be due to the nature of the components. The nanoemulsions and microemulsions can be easy to handle and formulate. They have low solubility rates in water so their application in the field is easy (Khater et al., 2017). The encapsulation capability of these molecules can enhance the stability of the pesticides (Perlatti et al., 2013). The nanoencapsulation of triazophos can be protected from hydrolysis (Nikos et al., 2017).

The nanosystems can ensure the release of the active ingredients to the target sites. The active compounds can easily spread, deposit, and can get permeated into the target sites. The essential oils can be used in both the flavor and fragrance industries. They can be used in the foodstuff. Most of the essential oils can be used as antimicrobial agents. The nanotechnology products can occupy most of the biopesticide markets (Isman et al., 2006). These substances can inflict low toxicity in mammals (Abbar et al., 2016). The pesticides or biopesticides can be sprayed onto the plants. The nanodispersion can be called as a nano tank that can deliver active ingredients. This can be attributed to the small size and so the penetration and mobility will be enhanced (Sasson et al., 2007). The essential oil-based nanosystems can be used as coarse emulsions. These essential oil-based nanoparticles can induce larvicidal activity in Vitex species. The encapsulated forms can be more effective than free counterparts. The microemulsions and nanoemulsions can exhibit a high rate of dispersion and so the active ingredients will be dispersed at the interface properly and so the targets will be killed (Gaysinsky et al., 2007). It was observed that the reduction in the drop size enhanced the efficacy of the formulations (Anjali et al., 2011). The reduction of size in the droplets has enhanced both penetration capacity and interaction rates. We can use the nanoemulsions and microemulsions to stop mites. Mites can produce much damage to the ornamental and vegetable plants in the greenhouses. We commonly use pesticides for this purpose.

8.7 ENVIRONMENTAL PROBLEMS AND CHALLENGES

In the early stages of nanotechnology-based materials, there were certain estimations that nanotechnology can pose a threat to the biotic components of the environment. Risk can be classified as an important issue in any technology. There are certain risks to both humans and the environment with nanotechnology-based products. The fate, toxicity, transport, and exposure of the nanomaterials can be classified as important issues. Ultrafine particles can be added to the atmosphere that can create problems with several environmental factors. Recently several studies were made with nano-engineered products. It is always important to consider the exposure, fate, and transport of these nanoparticles. The concentration of the nanoparticles can get diminished with distance from the point of application. It is still unclear to say that if the coated nanoparticles share the same fate as the nanoparticles used elsewhere. The release of carbon nanotubes in the environment can be of great concern as there will be dermal exposure. The nanoparticles of CeO_2 can be considered as an oxidant and they can polymerize and decarboxylate the organic molecules. The combustion mechanism can be enhanced with the addition of gasoline. This compound can get accumulated in organisms, water, and soils. The impact of these particles on the environment can be based on several environmental factors. They are biochemical reactions, pressure, heat, redox potential, and solution chemistry. The behavior and stability of these nanoparticles can also dictate the impacts. Most of the mass transfer studies pertaining to the nanoparticles were made and they were related to aerosol transport. The transport mechanisms were also studied in porous media. In this case, the components of the ecosystem were considered. Availability of oxygen, atmosphere, water, surface waters, oceans, sediments, and soils were taken into consideration.

Toxicity parameters were also studied in the nanoparticles. Due to increased surface area, the toxicity of the nanoparticles will also get increased. The application of fullerenes can cause organ damage, skin inflammation, and damage to the cells and proteins. Most of the carbon nanotubes are classified as cytotoxic compounds. These compounds can produce granulomas in the lungs. Some of the nanoparticles are metals and metal oxides. They can be toxic to the cells and in some cases, they can cause inflammation. The titanium dioxide nanoparticles can produce chromosomal aberrations and can cause DNA damage. The hydroxyapatite nanoparticles can have the same properties as the substances that are present in the teeth and bones. These nanoparticles can produce cell death. The semiconductor nanoparticles can induce toxicity on their exposure to light and surface derivatization. The fullerenes and nanoparticles can come under this category. The coatings of the nanoparticulates can be temporary and hence most of the catalytic properties, bioavailability, reactivity, solubility, and toxicity can be intact. These coatings may not persist for long periods. The quantum dots prepared from CdSe can be coated so that the cytotoxicity will be reduced. These compounds can be reactive and bioavailability will be high in the organisms. The CdSe quantum dots that are devoid of surface coatings can alter DNA and may be toxic to the cells. The impact of the nanoparticles on a global scale must be studied in detail. The properties of the clouds can be altered due to the release of the nanoparticulates in the atmosphere. The titanium dioxide can be used in the disinfection process. These nanoparticles can also degrade the pollutants. These nanoparticles can be used in organic transformations. This effect can be high if they are exposed to photochemical reactions. Most of the biogeochemical processes can be altered by the intervention of nanoparticles.

The environmental and energy challenges are important in the survival of human beings. These are essential in maintaining the sustainability of the planet. The latest technologies have provided many solutions that can address environmental remediation problems. The problems associated with renewable energy can be solved with these technologies. Environmental nanotechnology has become popular in several sectors. Since several studies pertaining to nanomaterials are in vogue, it is important to discuss the safety of these applications. The feasibility of these applications must be tested before their usage in the agricultural sectors. There should be a balance in the stability, selectivity, and activity of these nanoparticles.

Nanotechnology can be considered as the second industrial revolution. Any country with nanotechnology infrastructure can be classified as the technology capital of the world. There are several health hazards with the nanoparticles. Some of the nanoparticles can be classified as environmental toxins or poisons. They can get accumulated in the organs. Nanotechnology can be used in the form of systems, devices, and materials and this can be achieved by altering matter at the nanoscale. The supramolecular structures, molecules, and atoms can be modified at the nanoscale. Nanotechnology can be used in the protection of the environment. Green technologies can use nanotechnology to reduce environmental hazards. We can eliminate contaminants from water and air. So, we can mitigate the pollutants that are present in the environment. Nanotechnology can minimize waste and most of the energy will be conserved. Nanoscale materials can be used in environmental sensing systems. We can use nanomaterials in the form of distributed arrays of sensors. These sensors can be used in understanding the dynamics pertaining to biology, chemistry, and spatial-temporal variability. The nanoscale-based equipment are cheaper and consume less energy. There will be an improvement in the life cycle of the plants due to the application of nanoparticles. The nanowires can be electrochemically fabricated and it can be classified as a major advancement. We can alter the electrolytic solution so that the fabrication of the nanowires can be easy. The nanowires can be complex in their structure. The size and shape of these substances can provide us several advantages over conventional biological and chemical sensors. The cumulative effects of these materials can form the basis of biological and chemical molecules. Certain novel methods can be used in both the growth and assembly of these particles. With this technology, there will be the production of monolayer films.

The biotechnology revolution has brought about the latest needs for nanotechnology. Most of the nanotechnology applications can be originated from the crop sciences. There are several implications brought about by nanotechnology in global and natural processes. We can quantify the nanoparticles in almost every part of the ecosystem. Environmental sustainability can be regulated by nanotechnology. The nano-scale processes and nanoparticles can inflict an impact on climate change. Thermodynamically, the nanoscale structures can be better than macroscale structures.

Nanotechnology can regulate the controlling processes in agriculture. This is possible only with the reduced dimensions. There are many advantages with nanotechnology in the fields of agriculture and food industries. Natural resources, food, and agriculture can be part of this technology. Most of the nutrient losses can be reduced. There will be effective nutrient and pest management in agriculture. The limitations of agriculture can demand most of the latest technologies so that we can attempt novel techniques. The nutrient balances in agriculture can help in economic growth (Campbell et al., 2014). Nanotechnology can reduce the infections that come with crop yields (Yunlong et al., 1994). The nanoparticles can get involved in several food chain ecosystems. There are the latest instruments that can measure the nanomaterials and isolate them (Huey et al., 2007). The nanomaterials can be reactive and may be used in different biological processes (Prasad et al., 2017). There is a need for low-cost systems and reliable systems for the diagnosis of diseases in the agricultural sectors (Vidotti et al., 2011). Green technology has become popular these days due to its promising applications in agriculture. There are some risks with green nanotechnology also (Kandasamy et al., 2017). With the aid of nanotechnology, there will be precise management of the inputs in agriculture (Kah, 2014). The quantity and quality of the crops can be enhanced with the application of nanoparticles (Sekhon, 2014). We can build nanosensors to enhance precision farming practices (Fraceto et al., 2014). There are signal transduction pathways that rely on nanomaterials (Sertova, 2015). The nanomaterials can be used in their miniature forms in the form of nanosystems and nanosensors. In the food materials, we can detect the mycotoxins with the nanosensors. The quantum dots and nanoparticles can be used to label the plant proteins (Chahine et al., 2014). The carbon nanotubes can be used in many nanodevices. Agrochemicals can kill pests and they can be associated with carbon nanotubes. The gold nanorods can be used to bring about physiological changes in the watermelon plants (Wan et al., 2014). Precision targeting of the pests can be possible with the nanoencapsulation methods.

8.8 INDIRECT MEASUREMENT TECHNIQUES

The sensors can be used in testing the soils and it can be considered as an important aspect in knowing soil fertility. There are two different types of soil sorption zones wherein plants can grow easily. Most of the soil volumes are occupied by the roots. This can be called the root system sorption zone. The mobile nutrients can be absorbed by the plants. The sorption zone can be relatively thin and is closer to the root surface. This can be called the root surface sorption zone. At this zone, the immobile nutrients can be eliminated by the plants. The plants can respond to the mobile nutrients and immobile nutrients that are present in the soils. Most of the nutrition depletion conditions in the root system sorption zones can be based on growing conditions and environmental constraints. The crop yields can be proportional to the mobile nutrients that are present in the soil. The depletion of the nutrients in the root system sorption zone can be based on environmental conditions. The crop response to the immobile nutrients can be linked with the immobile nutrients.

Light can reach the surface of the soils and plants and most of the radiation is escaped via re-radiation, reflection, and other mechanisms. If the sensors detect yellow-green color, we can suspect that the plants are facing nitrogen deficiency. Most of the violet light can be absorbed by the plants in this case. The green plants can be identified by the absorption of the red light. When plants suffer from P deficiency, then we can get absorption maxima of green light. The chemical methods of investigating the nutrient deficiencies were replaced with non-destructive techniques. Spectral analyses can be classified as one of the techniques that can be classified as a safe method to know the nutritional status of the plants. Near-infrared spectroscopy was used to measure oil, fat, moisture, and protein in agricultural products. The leaf reflectance measurements can be used to know the nitrogen status of the sweet peppers. We can use the Near IR region to estimate the total nitrogen and organic carbon present in the soils. The organic compounds can be present special absorption properties in the NIR wavelengths. This can be observed in the form of bending and stretching vibrations that occur between the elements. One band can be sensitive to the amino acids and other bands can be sensitive to proteins.

The microwave polarization difference index can be used to estimate the water content in the plants. We can measure NDVI to know the chlorophyll absorption rates. The chlorophyll A absorption can be used to determine the availability of N in the plants. The physiological reflectance index can be used to know the photon efficiency. The spectral radiance was used to measure the soil characteristics along with the canopy properties. The sensor-based systems can lower the costs incurred whilst data management, chemical analysis, and soil sampling. The spectral reflectance data may be inversely proportional to the chlorophyll levels of the leaf. The optical spectra can be based on the pigments present in the leaves.

The absorption maxima are observed in the blue spectral region and red spectral regions. These regions can be observed in the chlorophyll band. The healthy plants can absorb red and blue light. Most of the infrared and green radiation can get reflected. In the case of chlorosis, the violet light is absorbed by the plants. In the case of P deficiency, most of the green light will be absorbed. Nitrogen deficiency can lead to reduced chlorophyll production so the same occurs with the spectral reflectance of the plant leaves. The chlorophyll rates can be an indicator of physiological stress. This can be the logic behind the detection of the plant stress conditions. Many remote sensing techniques can solve the problems linked to agriculture. Most of the problems can be associated with nutrient stress. Some of the sensors can be used to know water deficit, pests, and disease spread.

8.9 CONCLUSION

The nutrient use efficiency can estimate the crop yields in terms of biomass and grains in response to the nutrients applied. The crop plants can have elements that can play an important role in the growth and development of the crops. The genetically modified plants can enhance the nutrient use efficiency that can help attain high crop yields. Due to this, there will be enhanced bioavailability of both macroelements and microelements in the crops. Most of the areas pertaining to the improvement of plant growth yields were controlled by environmental factors. The vectors and pests can be classified as important as far as nanotechnology and agriculture are concerned. Most of the plants can respond to the shortage of nutrients and this can be reflected in the form of developmental, morphological, and physiological responses. Microarray technology has become popular these days to know the plant gene expression patterns. The gene patterns can be observed in diverse nutritional and environmental conditions. Genome-wide microarray analyses are being used in several fields. The growth processes of cells, gene regulation, protein synthesis, nutrient transport, primary and secondary metabolisms were studied using genome-wide microarray analyses. Most of the gene pools can respond to environmental stresses. Jefferson (1993) has formulated a criterion for this purpose. Green fluorescent protein or GFP was used in knowing the genetic expression that can be changed during imbalances in the environment (Tsien, 1998). The Gene fusion concept was also used to know the problems faced by the plants in nutrient deficiencies. The reporter genes can be used to evaluate the nutrient status of magnesium, sulfates, and phosphates (Beth et al., 2003). Transgenes were used to change the crop nutrient absorption rates (Rajesh et al., 2011). The growing crop varieties may need enhanced capacities to absorb nutrients and utilization and the study of these crops is essential. It is always important to choose the integrated nutrient management step before farming. It will help us to manage the nutrient deficiencies within the stipulated time.

References

1. Antoniadis, V., Robinson, J., & Alloway, B. (2008). Effects of short-term pH fluctuations on cadmium, nickel, lead, and zinc availability to ryegrass in a sewage sludge-amended field. Chemosphere,71(4), 759-764. doi:10.1016/j.chemosphere.2007.10.015
2. Arnon, D. I., & Stout, P. R. (1939, 07). Molybdenum as an Essential Element for Higher Plants. Plant Physiology,14(3), 599-602. doi:10.1104/pp.14.3.599
3. Arthur, J. R. (2003). Selenium supplementation: Does soil supplementation help and why? Proceedings of the Nutrition Society,62(2), 393-397. doi:10.1079/pns2003254
4. Ashworth, D., & Alloway, B. (2004, 01). Soil mobility of sewage sludge-derived dissolved organic matter, copper, nickel and zinc. Environmental Pollution,127(1), 137-144. doi:10.1016/s0269-7491(03)00237-9
5. Atkinson, J. A., Wingen, L. U., Griffiths, M., Pound, M. P., Gaju, O., Foulkes, (2015). Phenotyping pipeline reveals major seedling root growth QTL in hexaploid wheat. Journal of Experimental Botany,66(8), 2283-2292. doi:10.1093/jxb/erv006
6. Baish, J. W., Gazit, Y., Berk, D. A., Nozue, M., Baxter, L. T., & Jain, R. K. (1996). Role of Tumor Vascular Architecture in Nutrient and Drug Delivery: An Invasion Percolation-Based Network Model. Microvascular Research,51(3), 327-346. doi:10.1006/mvre.1996.0031
7. Baligar, V. C., Fageria, N. K., & He, Z. L. (2001, 04). Nutrient Use Efficiency In Plants. Communications in Soil Science and Plant Analysis,32(7-8), 921-950. doi:10.1081/css-100104098

8. Barber, S. A. (1995). Soil nutrient bioavailability: A mechanistic approach. New York: John Wiley Sons.
9. Barberon, M., Berthomieu, P., Clairotte, M., Shibagaki, N., Davidian, J., & Gosti, F. (2008). Unequal functional redundancy between the twoArabidopsis thaliana high-affinity sulphate transporters ULTR1;1andSULTR1;2. New Phytologist,180(3), 608-619. doi:10.1111/j.1469-8137.2008.02604.x
10. Barea, J. M., Navarro, E., & Montoya, E. (1976). Production of Plant Growth Regulators by Rhizosphere Phosphate-solubilizing Bacteria. Journal of Applied Bacteriology,40(2), 129-134. doi:10.1111/j.1365-2672.1976.tb04161.x
11. Barraclough, P. B., & Tinker, P. B. (1981). The Determination of Ionic Diffusion Coefficients In Field Soils. I. Diffusion Coefficients In Sieved Soils In Relation To Water Content And Bulk Density. Journal of Soil Science,32(2), 225-236. doi:10.1111/j.1365-2389.1981.tb01702.x
12. Bashan, Y., Kamnev, A. A., & De-Bashan, L. E. (2012). Tricalcium phosphate is inappropriate as a universal selection factor for isolating and testing phosphate-solubilizing bacteria that enhance plant growth: A proposal for an alternative procedure. Biology and Fertility of Soils,49(4), 465-479. doi:10.1007/s00374-012-0737-7
13. Bertsch, P. M., & Thomas, G. W. (2015, 11). Potassium Status of Temperate Region Soils. Potassium in Agriculture ASA, CSSA, and SSSA Books, 129-162. doi:10.2134/1985.potassium.c7
14. Boden, R., Hutt, L. P., & Rae, A. W. (2017). Reclassification of Thiobacillus aquaesulis (Wood & Kelly, 1995) as Annwoodia aquaesulis gen. nov., comb. nov., transfer of Thiobacillus (Beijerinck, 1904) from the Hydrogenophilales to the Nitrosomonadales, proposal of Hydrogenophilalia class. nov. within the 'Proteobacteria', and four new families within the orders Nitrosomonadales and Rhodocyclales. International Journal of Systematic and Evolutionary Microbiology,67(5), 1191-1205. doi:10.1099/ijsem.0.001927
15. Boiero, L., Perrig, D., Masciarelli, O., Penna, C., Cassán, F., & Luna, V. (2007). Phytohormone production by three strains of Bradyrhizobium japonicum and possible physiological and technological implications. Applied Microbiology and Biotechnology,74(4), 874-880. doi:10.1007/s00253-006-0731-9
16. Brady, C. L., Goszczynska, T., Venter, S. N., Cleenwerck, I., Vos, P. D., Gitaitis, R. D., & Coutinho, T. A. (2011, 04). Pantoea allii sp. nov.,

isolated from onion plants and seed. International Journal of Systematic and Evolutionary Microbiology,61(4), 932-937. doi:10.1099/ijs.0.022921-0

17. Broek, A. V., Gysegom, P., Ona, O., Hendrickx, N., Prinsen, E., Impe, J. V., & Vanderleyden, J. (2005). Transcriptional Analysis of the Azospirillum brasilense Indole-3-Pyruvate Decarboxylase Gene and Identification of a cis-Acting Sequence Involved in Auxin Responsive Expression. Molecular Plant-Microbe Interactions®,18(4), 311-323. doi:10.1094/mpmi-18-0311
18. Brun, F., Richard-Molard, C., Pagès, L., Chelle, M., & Ney, B. (2010). To what extent may changes in the root system architecture of Arabidopsis thaliana grown under contrasted homogenous nitrogen regimes be explained by changes in carbon supply? A modelling approach. Journal of Experimental Botany,61(8), 2157-2169. doi:10.1093/jxb/erq090
19. Bräutigam, A., & Weber, A. P. (2010). Do Metabolite Transport Processes Limit Photosynthesis? Plant Physiology,155(1), 43-48. doi:10.1104/pp.110.164970
20. Caplin, N., & Willey, N. (2018). Ionizing Radiation, Higher Plants, and Radioprotection: From Acute High Doses to Chronic Low Doses. Frontiers in Plant Science,9. doi:10.3389/fpls.2018.00847
21. Cathcart, J. B. (2015, 11). World Phosphate Reserves and Resources. The Role of Phosphorus in Agriculture ASA, CSSA, and SSSA Books, 1-18. doi:10.2134/1980.role of phosphorus.c1
22. Chandra, R., & Hoduck, K. (2018, 06). Phytoremediation and Physiological Effects of Mixed Heavy Metals on Poplar Hybrids. Heavy Metals. doi:10.5772/intechopen.76348
23. Claus, A., Weisz, G. M., Schieber, A., & Carle, R. (2006). Pyrolytic acrylamide formation from purified wheat gluten and gluten-supplemented wheat bread rolls. Molecular Nutrition & Food Research,50(1), 87-93. doi:10.1002/mnfr.200500152
24. Coque, M., Martin, A., Veyrieras, J. B., Hirel, B., & Gallais, A. (2008). Genetic variation for N-remobilization and postsilking N-uptake in a set of maize recombinant inbred lines. 3. QTL detection and coincidences. Theoretical and Applied Genetics,117(5), 729-747. doi:10.1007/s00122-008-0815-2
25. Deo, P. C., Harding, R. M., Taylor, M., Tyagi, A. P., & Becker, D. K. (2009). Somatic embryogenesis, organogenesis and plant regeneration

in taro (Colocasia esculenta var. esculenta). Plant Cell, Tissue and Organ Culture (PCTOC),99(1), 61-71. doi:10.1007/s11240-009-9576-0

26. Dersch, L. M., Beckers, V., & Wittmann, C. (2016). Green pathways: Metabolic network analysis of plant systems. Metabolic Engineering,34, 1-24. doi:10.1016/j.ymben.2015.12.001
27. Deubel, A., & Merbach, W. (2005). Influence of Microorganisms on Phosphorus Bioavailability in Soils. Soil Biology Microorganisms in Soils: Roles in Genesis and Functions, 177-191. doi:10.1007/3-540-26609-7_9
28. Dhungana, S., Anderson, D. S., Mietzner, T. A., & Crumbliss, A. L. (2005). Kinetics of Iron Release from Ferric Binding Protein (FbpA): Mechanistic Implications in Bacterial Periplasm-to-Cytosol Fe3 Transport†. Biochemistry,44(28), 9606-9618. doi:10.1021/bi0505518
29. Diaz, C., Lemaître, T., Christ, A., Azzopardi, M., Kato, Y., Sato, F., . . . Masclaux-Daubresse, C. (2008). Nitrogen Recycling and Remobilization Are Differentially Controlled by Leaf Senescence and Development Stage in Arabidopsis under Low Nitrogen Nutrition. Plant Physiology,147(3), 1437-1449. doi:10.1104/pp.108.119040
30. Dordas, C. (2006, 08). Foliar boron application affects lint and seed yield and improves seed quality of cotton grown on calcareous soils. Nutrient Cycling in Agroecosystems,76(1), 19-28. doi:10.1007/s10705-006-9037-7
31. Dunn, W. B., Bailey, N. J., & Johnson, H. E. (2005). Measuring the metabolome: Current analytical technologies. The Analyst,130(5), 606. doi:10.1039/b418288j
32. Dunn, W. B., Knowles, J. D., Broadhurst, D., Williams, R., Ashworth, J. J., Cameron, M., & Kell, D. B. (2007, 01). Closed-Loop, Multiobjective Optimization of Two-Dimensional Gas Chromatography/Mass Spectrometry for Serum Metabolomics. Analytical Chemistry,79(2), 464-476. doi:10.1021/ac061443
33. Eichert, T., & Fernández, V. (2012). Uptake and Release of Elements by Leaves and Other Aerial Plant Parts. Marschner's Mineral Nutrition of Higher Plants, 71-84. doi:10.1016/b978-0-12-384905-2.00004-2
34. Eldridge, D. J., Bowker, M. A., Maestre, F. T., Roger, E., Reynolds, J. F., & Whitford, W. G. (2011). Impacts of shrub encroachment on ecosystem structure and functioning: Towards a global synthesis. Ecology

Letters,14(7), 709-722. doi:10.1111/j.1461-0248.2011.01630.x

35. Erisman, J. W., Galloway, J. N., Seitzinger, S., Bleeker, A., Dise, N. B., Petrescu (2013). Consequences of human modification of the global nitrogen cycle. Philosophical Transactions of the Royal Society B: Biological Sciences,368(1621), 20130116. doi:10.1098/rstb.2013.0116
36. Esteller, M., Martínez-Valdés, H., Garrido, S., & Uribe, Q. (2009, 06). Nitrate and phosphate leaching in a Phaeozem soil treated with biosolids, composted biosolids and inorganic fertilizers. Waste Management,29(6), 1936-1944. doi:10.1016/j.wasman.2008.12.025
37. Fageria, N. K., Baligar, V. C., Moreira, A., & Portes, T. A. (2010, 10). Dry Bean Genotypes Evaluation For Growth, Yield Components And Phosphorus Use Efficiency. Journal of Plant Nutrition,33(14), 2167-2181. doi:10.1080/01904167.2010.519089
38. Fageria, N. K., Santos, A. B., & Coelho, A. M. (2011, 01). Growth, Yield And Yield Components Of Lowland Rice As Influenced By Ammonium Sulfate And Urea Fertilization. Journal of Plant Nutrition,34(3), 371-386. doi:10.1080/01904167.2011.536879
39. Fiehn, O. (2001). Combining Genomics, Metabolome Analysis, and Biochemical Modelling to Understand Metabolic Networks. Comparative and Functional Genomics,2(3), 155-168. doi:10.1002/cfg.82
40. Fischer, A. M. (2018). Nutrient Remobilization during Leaf Senescence. Annual Plant Reviews Online, 87-107. doi:10.1002/9781119312994.apr0266
41. Fischer, G., Winiwarter, W., Ermolieva, T., Cao, G., Qui, H., Klimont (2010). Integrated modeling framework for assessment and mitigation of nitrogen pollution from agriculture: Concept and case study for China. Agriculture, Ecosystems & Environment,136(1-2), 116-124. doi:10.1016/j.agee.2009.12.004
42. Fiset, C., Irwin, A. J., & Finkel, Z. V. (2019). The macromolecular composition of noncalcified marine macroalgae. Journal of Phycology,55(6), 1361-1369. doi:10.1111/jpy.12913
43. Flint-Garcia, S. A. (2013). Genetics and Consequences of Crop Domestication. Journal of Agricultural and Food Chemistry,61(35), 8267-8276. doi:10.1021/jf305511d
44. Follett, R., & Walker, D. (1989). Ground Water Quality Concerns About Nitrogen. Nitrogen Management and Ground Water Protection

Developments in Agricultural and Managed Forest Ecology, 1-22. doi:10.1016/b978-0-444-87393-4.50007-1

45. Forieri, I., Sticht, C., Reichelt, M., Gretz, N., Hawkesford, M. J., Malagoli (2016). System analysis of metabolism and the transcriptome in Arabidopsis thaliana roots reveals differential co-regulation upon iron, sulfur and potassium deficiency. Plant, Cell & Environment,40(1), 95-107. doi:10.1111/pce.12842
46. Freney, J. R. (2015). Forms and Reactions of Organic Sulfur Compounds in Soils. Agronomy Monographs Sulfur in Agriculture, 207-232. doi:10.2134/agronmonogr27.c6
47. Gaither, L. A., & Eide, D. J. (2001). Eukaryotic zinc transporters and their regulation. Zinc Biochemistry, Physiology, and Homeostasis, 65-84. doi:10.1007/978-94-017-3728-9_5
48. George, T. S., Simpson, R. J., Hadobas, P. A., Marshall, D. J., & Richardson, A. E. (2007). Accumulation and phosphatase-lability of organic phosphorus in fertilised pasture soils. Australian Journal of Agricultural Research,58(1), 47. doi:10.1071/ar06167
49. Giege, P., Sweetlove, L. J., & Leaver, C. J. (2003). Identification of mitochondrial protein complexes inArabidopsis using two-dimensional blue-native polyacrylamide gel electrophoresis. Plant Molecular Biology Reporter,21(2), 133-144. doi:10.1007/bf02774240
50. Grichko, V. P., & Glick, B. R. (2001). Amelioration of flooding stress by ACC deaminase-containing plant growth-promoting bacteria. Plant Physiology and Biochemistry,39(1), 11-17. doi:10.1016/s0981-9428(00)01212-2
51. Guo, T., Wei, L., Sun, J., Hou, C., & Fan, L. (2011). Antioxidant activities of extract and fractions from Tuber indicum Cooke & Massee. Food Chemistry,127(4), 1634-1640. doi:10.1016/j.foodchem.2011.02.030
52. Hancock, A. M., Brachi, B., Faure, N., Horton, M. W., Jarymowycz, L. B., Sperone, Bergelson, J. (2011). Adaptation to Climate Across the Arabidopsis thaliana Genome. Science,334(6052), 83-86. doi:10.1126/science.1209244
53. Hargreaves, S. K., Horrigan, E. J., & Jefferies, R. L. (2009). Seasonal partitioning of resource use and constraints on the growth of soil microbes and a forage grass in a grazed Arctic salt-marsh. Plant and Soil,322(1-2), 279-291. doi:10.1007/s11104-009-9915-2

54. Haruna, S. I., & Nkongolo, N. V. (2020, 06). Influence of Cover Crop, Tillage, and Crop Rotation Management on Soil Nutrients. Agriculture,10(6), 225. doi:10.3390/agriculture10060225

55. He, J., Li, H., Rasaily, R. G., Wang, Q., Cai, G., Su, Y (2011). Soil properties and crop yields after 11 years of no tillage farming in wheat–maize cropping system in North China Plain. Soil and Tillage Research,113(1), 48-54. doi:10.1016/j.still.2011.01.005

56. Hewitt, E. J., & Bolle-Jones, E. W. (1952, 01). Molybdenum as a Plant Nutrient: II. The Effects of Molybdenum Deficiency on Some Horticultural and Agricultural Crop Plants in Sand Culture. Journal of Horticultural Science,27(4), 257-265. doi:10.1080/00221589.1952.11513763

57. Holme, P., Huss, M., & Jeong, H. (2003). Subnetwork hierarchies of biochemical pathways. Bioinformatics,19(4), 532-538. doi:10.1093/bioinformatics/btg033

58. Howarth, J. R., Fourcroy, P., Davidian, J., Smith, F. W., & Hawkesford, M. J. (2003). Cloning of two contrasting high-affinity sulfate transporters from tomato induced by low sulfate and infection by the vascular pathogen Verticillium dahliae. Planta,218(1), 58-64. doi:10.1007/s00425-003-1085-5

59. Huang, X., Paulo, M., Boer, M., Effgen, S., Keizer, P., Koornneef, M., & Eeuwijk, F. A. (2011). Analysis of natural allelic variation in Arabidopsis using a multiparent recombinant inbred line population. Proceedings of the National Academy of Sciences,108(11), 4488-4493. doi:10.1073/pnas.1100465108

60. Hust, B., & Gutensohn, M. (2006). Deletion of Core Components of the Plastid Protein Import Machinery Causes Differential Arrest of Embryo Development in Arabidopsis thaliana. Plant Biology,8(1), 18-30. doi:10.1055/s-2005-873044

61. Islam, M. T., Deora, A., Hashidoko, Y., Rahman, A., Ito, T., & Tahara, S. (2007). Isolation and Identification of Potential Phosphate Solubilizing Bacteria from the Rhizoplane of Oryza sativa L. cv. BR29 of Bangladesh. Zeitschrift Für Naturforschung C,62(1-2), 103-110. doi:10.1515/znc-2007-1-218

62. Janosch, C., Thyssen, C., Vera, M. A., Bonnefoy, V., Rohwerder, T., & Sand, W. (2009). Sulfur Oxygenase Reductase in Different Acidithio bacillus Caldus-Like Strains. Advanced Materials Research,71-73, 239-242. doi:10.4028/www.scientific.net/amr.71-73.239

63. Jeon, J., Lee, S., Jung, K., Jun, S., Jeong G. (2000, 06). T-DNA insertional mutagenesis for functional genomics in rice. The Plant Journal,22(6), 561-570. doi:10.1046/j.1365-313x.2000.00767.x
64. Johansson, E., Malik, A. H., Hussain, A., Rasheed, F., Newson, W. R., Plivelic (2013). Wheat Gluten Polymer Structures: The Impact of Genotype, Environment, and Processing on Their Functionality in Various Applications. Cereal Chemistry Journal,90(4), 367-376. doi:10.1094/cchem-08-12-0105-fi
65. Johnson, G. V., & Fixen, P. E. (2018, 09). Testing Soils for Sulfur, Boron, Molybdenum, and Chlorine. SSSA Book Series Soil Testing and Plant Analysis, 265-273. doi:10.2136/sssabookser3.3ed.c10
66. Jolley, K. A., & Maiden, M. C. (2010). BIGSdb: Scalable analysis of bacterial genome variation at the population level. BMC Bioinformatics,11(1). doi:10.1186/1471-2105-11-595
67. Kang, B. G., Kim, W. T., Yun, H. S., & Chang, S. C. (2010). Use of plant growth-promoting rhizobacteria to control stress responses of plant roots. Plant Biotechnology Reports,4(3), 179-183. doi:10.1007/s11816-010-0136-1
68. Kantachote, D., Nunkaew, T., Kantha, T., & Chaiprapat, S. (2016). Biofertilizers from Rhodopseudomonas palustris strains to enhance rice yields and reduce methane emissions. Applied Soil Ecology,100, 154-161. doi:10.1016/j.apsoil.2015.12.015
69. Kataoka, T., Watanabe-Takahashi, A., Hayashi, N., Ohnishi, M., Mimura, T., Buchner, H. (2004). Vacuolar Sulfate Transporters Are Essential Determinants Controlling Internal Distribution of Sulfate in Arabidopsis . The Plant Cell,16(10), 2693-2704. doi:10.1105/tpc.104.023960
70. Kellermeier, F., Chardon, F., & Amtmann, A. (2013). Natural Variation of Arabidopsis Root Architecture Reveals Complementing Adaptive Strategies to Potassium Starvation. Plant Physiology,161(3), 1421-1432. doi:10.1104/pp.112.211144
71. Kende, H. (1989). Enzymes of Ethylene Biosynthesis. Plant Physiology,91(1), 1-4. doi:10.1104/pp.91.1.1
72. Keren, R., & Bingham, F. T. (1958). Boron in Water, Soils, and Plants. Advances in Soil Science, 229-276. doi:10.1007/978-1-4612-5046-3_7

73. Kotera, M., Tokimatsu, T., Kanehisa, M., & Goto, S. (2011). MUCHA: Multiple chemical alignment algorithms to identify building block substructures of orphan secondary metabolites. BMC

74. Krishnan, P. (2004, 12). Metabolite fingerprinting and profiling in plants using NMR. Journal of Experimental Botany,56(410), 255-265. doi:10.1093/jxb/eri010

75. Lal, R., & Logan, T. (2018). Agricultural Activities and Greenhouse Gas Emissions from Soils of the Tropics. Soil Management and Greenhouse Effect, 293-308. doi:10.1201/9780203739310-25

76. Li, L., Coppola, E., Rine, J., Miller, J. L., & Walker, D. (2010). Catalytic Hydrothermal Conversion of Triglycerides to Non-ester Biofuels. Energy & Fuels,24(2), 1305-1315. doi:10.1021/ef901163a

77. Liberman, L. M., Sozzani, R., & Benfey, P. N. (2012). Integrative systems biology: An attempt to describe a simple weed. Current Opinion in Plant Biology,15(2), 162-167. doi:10.1016/j.pbi.2012.01.004

78. Lorence, A., & Nessler, C. L. (2007). Pathway Engineering of the Plant Vitamin C Metabolic Network. Applications of Plant Metabolic Engineering, 197-217. doi:10.1007/978-1-4020-6031-1_8

79. Lu, K., Knutson, C. G., Wishnok, J. S., Fox, J. G., & Tannenbaum, S. R. (2012). Serum Metabolomics in a Helicobacter hepaticusMouse Model of Inflammatory Bowel Disease Reveal Important Changes in the Microbiome, Serum Peptides, and Intermediary Metabolism. Journal of Proteome Research,11(10), 4916-4926. doi:10.1021/pr300429x

80. Macmillan, J. (2002). Erratum: Occurrence of Gibberellins in Vascular Plants, Fungi, and Bacteria. Journal of Plant Growth Regulation,21(3), 242-243. doi:10.1007/s00344-003-0004-0

81. Macnicol, R. D., & Beckett, P. H. (1985). Critical tissue concentrations of potentially toxic elements. Plant and Soil,85(1), 107-129. doi:10.1007/bf02197805

82. Mahon, J. D. (1983, 01). Limitations to The Use of Physiological Variability in Plant Breeding. Canadian Journal of Plant Science,63(1), 11-21. doi:10.4141/cjps83-002

83. Marschner, H. (1993). Zinc Uptake from Soils. Zinc in Soils and Plants, 59-77. doi:10.1007/978-94-011-0878-2_5

84. Maruyama-Nakashita, A., Nakamura, Y., Yamaya, T., & Takahashi, H. (2004). A novel regulatory pathway of sulfate uptake inArabidopsis roots: Implication of CRE1/WOL/AHK4-mediated cytokinin-

dependent regulation. The Plant Journal,38(5), 779-789. doi:10.1111/j.1365-313x.2004.02079.x

85. Mengel, K., Kirkby, E. A., Kosegarten, H., & Appel, T. (2001). Molybdenum. Principles of Plant Nutrition, 613-619. doi:10.1007/978-94-010-1009-2_17
86. Michalczyk, D., Drozdowicz, A., Pintscher, S., & Plonka, P. (2011). Mycetozoal bloom in hydroponic culture of garden cress (Lepidium sativum L.). International Journal of Food Microbiology,145(1), 336-341. doi:10.1016/j.ijfoodmicro.2010.11.042
87. Miflin, B. (2000). Crop improvement in the 21st century. Journal of Experimental Botany,51(342), 1-8. doi:10.1093/jexbot/51.342.1
88. Miyata, K., Kozaki, T., Kouzai, Y., Ozawa, K., Ishii, K., Asamizu, E, Nakagawa, T. (2014, 09). The Bifunctional Plant Receptor, OsCERK1, Regulates Both Chitin-Triggered Immunity and Arbuscular Mycorrhizal Symbiosis in Rice. Plant and Cell Physiology,55(11), 1864-1872. doi:10.1093/pcp/pcu129
89. Moco, S., Bino, R. J., Vorst, O., Verhoeven, H. A., Groot, J. D., Beek (2006). A Liquid Chromatography-Mass Spectrometry-Based Metabolome Database for Tomato. Plant Physiology,141(4), 1205-1218. doi:10.1104/pp.106.078428
90. Monneveux, P., Zaidi, P. H., & Sanchez, C. (2005). Population Density and Low Nitrogen Affects Yield-Associated Traits in Tropical Maize. Crop Science,45(2), 535-545. doi:10.2135/cropsci2005.0535
91. Moulin, M., Mccormac, A. C., Terry, M. J., & Smith, A. G. (2008). Tetrapyrrole profiling in Arabidopsis seedlings reveals that retrograde plastid nuclear signaling is not due to Mg-protoporphyrin IX accumulation. Proceedings of the National Academy of Sciences,105(39), 15178-15183. doi:10.1073/pnas.0803054105
92. Murillo, M., & Leustek, T. (1995). Adenosine-5′-Triphosphate-Sulfurylase from Arabidopsis thaliana and Escherichia coli Are Functionally Equivalent but Structurally and Kinetically Divergent: Nucleotide Sequence of Two Adenosine-5′-Triphosphate-Sulfurylase cDNAs from Arabidopsis thaliana and Analysis of a Recombinant Enzyme. Archives of Biochemistry and Biophysics,323(1), 195-204. doi:10.1006/abbi.1995.0026
93. Musetti, R., Farhan, K., Marco, F. D., Polizzotto, R., Paolacci, A., Ciaffi (2012). Differentially-regulated defence genes in Malus domestica

during phytoplasma infection and recovery. European Journal of Plant Pathology,136(1), 13-19. doi:10.1007/s10658-012-0147-6

94. Nambara, E., & Marion-Poll, A. (2003). ABA action and interactions in seeds. Trends in Plant Science,8(5), 213-217. doi:10.1016/s1360-1385(03)00060-8
95. Neumann, P. M. (2007). Evidence for long-distance xylem transport of signal peptide activity from tomato roots. Journal of Experimental Botany,58(8), 2217-2223. doi:10.1093/jxb/erm081
96. Neumann, R. (2019). Methane Oxidation in the Rhizosphere of Wetland Plants. Final Report. doi:10.2172/1573358
97. O'Brien, T., Beall, F. D., & Smith, H. (1985). De-Etiolation and Plant Hormones. Hormonal Regulation of Development III, 282-307. doi:10.1007/978-3-642-67734-2_9
98. OECD-FAO Agricultural Outlook 2000. (2000, 04). OECD-FAO Agricultural Outlook. doi:10.1787/agr_outlook-2000-en
99. OECD-FAO Agricultural Outlook 2003. (2003, 07). OECD-FAO Agricultural Outlook. doi:10.1787/agr_outlook-2003-en
100. Oliver, D. J., & Raman, R. (1995). Glycine decarboxylase: Protein chemistry and molecular biology of the major protein in leaf mitochondria. Journal of Bioenergetics and Biomembranes,27(4), 407-414. doi:10.1007/bf02110003
101. Osorio, N. W., & Habte, M. (2012). Phosphate desorption from the surface of soil mineral particles by a phosphate-solubilizing fungus. Biology and Fertility of Soils,49(4), 481-486. doi:10.1007/s00374-012-0763-5
102. Page, V., Weisskopf, L., & Feller, U. (2006). Heavy metals in white lupin: Uptake, root-to-shoot transfer and redistribution within the plant. New Phytologist,171(2), 329-341. doi:10.1111/j.1469-8137.2006.01756.x
103. Parkin, T. B., & Meisinger, J. J. (1989, 01). Denitrification below the Crop Rooting Zone as Influenced by Surface Tillage. Journal of Environmental Quality,18(1), 12-16. doi:10.2134/jeq1989.00472425001800010002x
104. Pegler, R. A. (1976, 03). Harvest ripeness in grass seed crops. Grass and Forage Science,31(1), 7-13. doi:10.1111/j.1365-2494.1976.tb01108.x
105. Pennisi, E. (2009). Stressed Out Over a Stress Hormone. Science,324(5930), 1012-1013. doi:10.1126/science.324_1012

106. Pichersky, E., & Lewinsohn, E. (2011). Convergent Evolution in Plant Specialized Metabolism. Annual Review of Plant Biology,62(1), 549-566. doi:10.1146/annurev-arplant-042110-103814
107. Pollmann, S., Neu, D., & Weiler, E. W. (2003). Molecular cloning and characterization of an amidase from Arabidopsis thaliana capable of converting indole-3-acetamide into the plant growth hormone, indole-3-acetic acid. Phytochemistry,62(3), 293-300. doi:10.1016/s0031-9422(02)00563-0
108. Poolman, M. G., Miguet, L., Sweetlove, L. J., & Fell, D. A. (2009). A Genome-Scale Metabolic Model of Arabidopsis and Some of Its Properties. Plant Physiology,151(3), 1570-1581. doi:10.1104/pp.109.141267
109. Radzki, W., Mañero, F. J., Algar, E., García, J. A., García-Villaraco, A., & Solano, B. R. (2013). Bacterial siderophores efficiently provide iron to iron-starved tomato plants in hydroponics culture. Antonie Van Leeuwenhoek,104(3), 321-330. doi:10.1007/s10482-013-9954-9
110. Rae, A., & Smith, F. (2002). Localisation of expression of a high-affinity sulfate transporter in barley roots. Planta,215(4), 565-568. doi:10.1007/s00425-002-0777-6
111. Rafalski, J. A. (2010). Association genetics in crop improvement. Current Opinion in Plant Biology,13(2), 174-180. doi:10.1016/j.pbi.2009.12.004
112. Raun, W. R., & Johnson, G. V. (1999, 05). Improving Nitrogen Use Efficiency for Cereal Production. Agronomy Journal,91(3), 357-363. doi:10.2134/agronj1999.00021962009100030001x
113. Raymond, K. N., Dertz, E. A., & Kim, S. S. (2003). Enterobactin: An archetype for microbial iron transport. Proceedings of the National Academy of Sciences,100(7), 3584-3588. doi:10.1073/pnas.0630018100
114. Reich, M., Aghajanzadeh, T., Stuiver, C. E., Koralewska, A., & Kok, L. J. (2015). Impact of Sulfate Salinity on the Uptake and Metabolism of Sulfur in Chinese Cabbage. Proceedings of the International Plant Sulfur Workshop Molecular Physiology and Ecophysiology of Sulfur, 227-238. doi:10.1007/978-3-319-20137-5_25
115. Ribaut, J., Vicente, M. D., & Delannay, X. (2010). Molecular breeding in developing countries: Challenges and perspectives. Current Opinion in Plant Biology,13(2), 213-218. doi:10.1016/j.pbi.2009.12.011

116. Rose, T. J., Pariasca-Tanaka, J., Rose, M. T., Fukuta, Y., & Wissuwa, M. (2010). Genotypic variation in grain phosphorus concentration, and opportunities to improve P-use efficiency in rice. Field Crops Research,119(1), 154-160. doi:10.1016/j.fcr.2010.07.004

117. Rotte, C., & Leustek, T. (2000). Differential Subcellular Localization and Expression of ATP Sulfurylase and 5′-Adenylylsulfate Reductase during Ontogenesis of Arabidopsis Leaves Indicates That Cytosolic and Plastid Forms of ATP Sulfurylase May Have Specialized Functions. Plant Physiology,124(2), 715-724. doi:10.1104/pp.124.2.715

118. Russell, H. F. (1985). The Historical Geography of the Euphrates and Habur According to the Middle- and Neo-Assyrian Sources. Iraq,47, 57. doi:10.2307/4200232

119. Sabu, M., & Kuttan, R. (2002). Anti-diabetic activity of medicinal plants and their relationship with their antioxidant property. Journal of Ethnopharmacology,81(2), 155-160. doi:10.1016/s0378-8741(02)00034-x

120. Sakakibara, H. (2006). CYTOKININS: Activity, Biosynthesis, and Translocation. Annual Review of Plant Biology,57(1), 431-449. doi:10.1146/annurev.arplant.57.032905.105231

121. Salvi, S., Sponza, G., Morgante, M., Tomes, D., Niu, X., Fengler, Tuberosa, R. (2007). Conserved noncoding genomic sequences associated with a flowering-time quantitative trait locus in maize. Proceedings of the National Academy of Sciences,104(27), 11376-11381. doi:10.1073/pnas.0704145104

122. Sapkota, T., Aryal, J., Thapa, S., & Karki, A. B. (2013, 03). Biogas Production from Anaerobic Digestion of Different Biodegradable Materials. Nepal Journal of Science and Technology,13(2), 123-128. doi:10.3126/njst.v13i2.7724

123. Saravanan, V., Madhaiyan, M., & Thangaraju, M. (2007). Solubilization of zinc compounds by the diazotrophic, plant growth-promoting bacterium Gluconacetobacter diazotrophicus. Chemosphere,66(9), 1794-1798. doi:10.1016/j.chemosphere.2006.07.067

124. Schmidt, J., Seiler, W., & Conrad, R. (1988). Emission of nitrous oxide from temperate forest soils into the atmosphere. Journal of Atmospheric Chemistry,6(1-2), 95-115. doi:10.1007/bf00048334

125. Shaddox, T. W., Unruh, J. B., & Trenholm, L. E. (2017). Nitrogen Required for Acceptable Centipedegrass Quality, Color, Growth Rate, and Nitrate Leaching. Itsrj,13(1), 86. doi:10.2134/itsrj2016.04.0241

126. Sheldrick, W. (2015, 11). World Potassium Reserves. Potassium in Agriculture ASA, CSSA, and SSSA Books, 1-28. doi:10.2134/1985.potassium.c1

127. Shen, J., Tang, A., Liu, X., Fangmeier, A., Goulding, K., & Zhang, F. (2009). High concentrations and dry deposition of reactive nitrogen species at two sites in the North China Plain. Environmental Pollution,157(11), 3106-3113. doi:10.1016/j.envpol.2009.05.016

128. Shewry, P. R., Tatham, A. S., & Halford, N. G. (2001). Plant Growth Regulation,34(1), 105-111. doi:10.1023/a:1013382803849

129. Sivakumar, T., Ambika, S., & Balakrishnan, K. (2017). Biopriming of rice seed with phosphobacteria for enhanced germination and vigour. ORYZA- An International Journal on Rice,54(3), 346. doi:10.5958/2249-5266.2017.00048.0

130. Smil, V. (1999, 06). Nitrogen in crop production: An account of global flows. Global Biogeochemical Cycles,13(2), 647-662. doi:10.1029/1999gb900015

131. Spirin, V., & Mirny, L. A. (2003). Protein complexes and functional modules in molecular networks. Proceedings of the National Academy of Sciences,100(21), 12123-12128. doi:10.1073/pnas.2032324100

132. Sørhagen, K., Laxa, M., Peterhänsel, C., & Reumann, S. (2013). The emerging role of photorespiration and non-photorespiratory peroxisomal metabolism in pathogen defence. Plant Biology,15(4), 723-736. doi:10.1111/j.1438-8677.2012.00723.x

133. Staats, H., Jansen, L., & Thøgersen, J. (2011). Greening the greenhouse grower. A behavioral analysis of a sector-initiated system to reduce the environmental load of greenhouses. Journal of Environmental Management,92(10), 2461-2469. doi:10.1016/j.jenvman.2011.05.007

134. Stevenson, F. J., & Fitch, A. (2015, 10). Chemistry of Complexation of Metal Ions with Soil Solution Organics. SSSA Special Publications Interactions of Soil Minerals with Natural Organics and Microbes, 29-58. doi:10.2136/sssaspecpub17.c2

135. Stewart, W. M., Dibb, D. W., Johnston, A. E., & Smyth, T. J. (2005, 01). The Contribution of Commercial Fertilizer Nutrients to Food Production. Agronomy Journal,97(1), 1-6. doi:10.2134/agronj2005.0001

136. Stroud, J. L., Li, H. F., Lopez-Bellido, F. J., Broadley, M. R., Foot, I., Fairweather-Tait, Zhao, F. J. (2009). Impact of sulfur fertilization on

crop response to selenium fertilization. Plant and Soil,332(1-2), 31-40. doi:10.1007/s11104-009-0230-8

137. Tabuchi, M., Abiko, T., & Yamaya, T. (2007). Assimilation of ammonium ions and reutilization of nitrogen in rice (Oryza sativa L.). Journal of Experimental Botany,58(9), 2319-2327. doi:10.1093/jxb/erm016

138. Takahashi, H., Kopriva, S., Giordano, M., Saito, K., & Hell, R. (2011). Sulfur Assimilation in Photosynthetic Organisms: Molecular Functions and Regulations of Transporters and Assimilatory Enzymes. Annual Review of Plant Biology,62(1), 157-184. doi:10.1146/annurev-arplant-042110-103921

139. Tang, X., Liu, S., Liu, J., & Zhou, G. (2009). Effects of vegetation restoration and slope positions on soil aggregation and soil carbon accumulation on heavily eroded tropical land of Southern China. Journal of Soils and Sediments,10(3), 505-513. doi:10.1007/s11368-009-0122-9

140. Tisdale, S. L., Reneau, R. B., &Plantou, J. S. (2015, 10). Atlas of Sulfur Deficiencies. Agronomy Monographs Sulfur in Agriculture, 295-322. doi:10.2134/agronmonogr27.c10

141. Tiwari, K. N., Tiwari, V., Singh, J., Singh, B. D., & Ahuja, P. (2012). Synergistic effect of trimethoprim and bavistin for micropropagation of Bacopa monniera. Biologia Plantarum,56(1), 177-180. doi:10.1007/s10535-012-0038-x

142. Tseng, M., Chiu, (. S., Tan, R. R., & Siriban-Manalang, A. B. (2013). Sustainable consumption and production for Asia: Sustainability through green design and practice. Journal of Cleaner Production,40, 1-5. doi:10.1016/j.jclepro.2012.07.015

143. Turnbull, C. G., Crozier, A., Schwenen, L., & Graebe, J. E. (1985). Conversion of [14C] gibberellin A12-aldehyde to C19- and C20-gibberellins in a cell-free system from the immature seed of Phaseolus coccineus L. Planta,165(1), 108-113. doi:10.1007/bf00392218

144. Vazquez, A., Dobrin, R., Sergi, D., Eckmann, J. P., Oltvai, Z. N., & Barabasi, A. L. (2004). The topological relationship between the large-scale attributes and local interaction patterns of complex networks. Proceedings of the National Academy of Sciences,101(52), 17940-17945. doi:10.1073/pnas.0406024101

145. Voznesenskaya, E. V., Franceschi, V. R., Kiirats, O., Freitag, H., & Edwards, G. E. (2001). Kranz anatomy is not essential for terrestrial C4 plant photosynthesis. Nature,414(6863), 543-546. doi:10.1038/35107073
146. Vries, F. T., Thebault, E., Liiri, M., Birkhofer, K., Tsiafouli, M. A., Bjornlund, L., Bardgett, R. D. (2013, 08). Soil food web properties explain ecosystem services across European land-use systems. Proceedings of the National Academy of Sciences,110(35), 14296-14301. doi:10.1073/pnas.1305198110
147. Weightman, R. M., Millar, S., Alava, J., Foulkes, M. J., Fish, L., & Snape, J. W. (2008). Effects of drought and the presence of the 1BL/1RS translocation on grain rtu, hardness, and protein content in winter wheat. Journal of Cereal Science,47(3), 457-468. doi:10.1016/j.jcs.2007.05.011
148. White, C. N., Proebsting, W. M., Hedden, P., & Rivin, C. J. (2000). Gibberellins and Seed Development in Maize. I. Evidence That Gibberellin/Abscisic Acid Balance Governs Germination versus Maturation Pathways. Plant Physiology,122(4), 1081-1088. doi:10.1104/pp.122.4.1081
149. Wiersma, J. V. (2005, 05). High Rates of Fe-EDDHA and Seed Iron Concentration Suggest Partial Solutions to Iron Deficiency in Soybean. Agronomy Journal,97(3), 924-934. doi:10.2134/agronj2004.0309
150. Williams, T. C., Poolman, M. G., Howden, A. J., Schwarzlander, M., Fell, D. A., Ratcliffe, R. G., & Sweetlove, L. J. (2010). A Genome-Scale Metabolic Model Accurately Predicts Fluxes in Central Carbon Metabolism under Stress Conditions. Plant Physiology,154(1), 311-323. doi:10.1104/pp.110.158535
151. Wuchty, S., & Almaas, E. (2005). Peeling the yeast protein network. Proteomics,5(2), 444-449. doi:10.1002/pmic.200400962
152. Yamori, W., Takahashi, S., Makino, A., Price, G. D., Badger, M. R., & Caemmerer, S. V. (2010). The Roles of ATP Synthase and the Cytochrome b6/f Complexes in Limiting Chloroplast Electron Transport and Determining Photosynthetic Capacity. Plant Physiology,155(2), 956-962. doi:10.1104/pp.110.168435
153. Yang, J., & Zhang, J. (2005). Grain filling of cereals under soil drying. New Phytologist,169(2), 223-236. doi:10.1111/j.1469-8137.2005.01597.x

154. Yogendra, K. N., Pushpa, D., Mosa, K. A., Kushalappa, A. C., Murphy, A., & Mosquera, T. (2014). Quantitative resistance in potato leaves to late blight associated with induced hydroxycinnamic acid amides. Functional & Integrative Genomics,14(2), 285-298. doi:10.1007/s10142-013-0358-8
155. Yoshimoto, N., Inoue, E., Saito, K., Yamaya, T., & Takahashi, H. (2003). Phloem-Localizing Sulfate Transporter, Sultr1;3, Mediates Re-Distribution of Sulfur from Source to Sink Organs in Arabidopsis. Plant Physiology,131(4), 1511-1517. doi:10.1104/pp.014712
156. Zhao, F., Wood, A., & Mcgrath, S. (1999). Plant and Soil,212(2), 207-217. doi:10.1023/a:1004618303445
157. Abdel-Aziz, H. M., Hasaneen, M. N., & Omer, A. M. (2016, 03). Nano chitosan-NPK fertilizer enhances the growth and productivity of wheat plants grown in sandy soil. Spanish Journal of Agricultural Research,14(1). doi:10.5424/sjar/2016141-8205
158. Cox, A., Venkatachalam, P., Sahi, S., & Sharma, N. (2016, 10). Silver and titanium dioxide nanoparticle toxicity in plants: A review of current research. Plant Physiology and Biochemistry,107, 147-163. doi:10.1016/j.plaphy.2016.05.022
159. Davarpanah, S., Tehranifar, A., Davarynejad, G., Abadía, J., & Khorasani, R. (2016, 10). Effects of foliar applications of zinc and boron nano-fertilizers on pomegranate (Punica granatum cv. Ardestani) fruit yield and quality. Scientia Horticulturae,210, 57-64. doi:10.1016/j.scienta.2016.07.003
160. Disfani, M. N., Mikhak, A., Kassaee, M. Z., & Maghari, A. (2016, 11). Effects of nano Fe/SiO2fertilizers on germination and growth of barley and maize. Archives of Agronomy and Soil Science,63(6), 817-826. do i:10.1080/03650340.2016.1239016
161. Dobhal, N., & Raghuvanshi, R. S. (2018, 03). Physical characteristics and effect of germination on functional properties of black soybean (Glycine max). Asian Journal of Dairy and Food Research, (00). doi:10.18805/ajdfr.dr-1320
162. Dubey, A., & Mailapalli, D. R. (2016). sNano fertilizers, Nanopesticides, Nanosensors of Pest and Nanotoxicity in Agriculture. Sustainable Agriculture Reviews, 307-330. doi:10.1007/978-3-319-26777-7_7
163. Gao, J., Wang, Y., Folta, K. M., Krishna, V., Bai, W., Indeglia, P., . . . Moudgil, B. (2011, 05). Polyhydroxy Fullerenes (Fullerols

or Fullerenols): Beneficial Effects on Growth and Lifespan in Diverse Biological Models. PLoS ONE,6(5). doi:10.1371/journal.pone.0019976

164. Hou, R., Zhang, Z., Pang, S., Yang, T., Clark, J. M., & He, L. (2016, 06). Alteration of the Nonsystemic Behavior of the Pesticide Ferbam on Tea Leaves by Engineered Gold Nanoparticles. Environmental Science & Technology,50(12), 6216-6223. doi:10.1021/acs.est.6b01336
165. Joshi, A., Kaur, S., Dharamvir, K., Nayyar, H., & Verma, G. (2018, 01). Multi-walled carbon nanotubes applied through seed-priming influence early germination, root hair, growth, and yield of bread wheat (Triticum aestivum L.). Journal of the Science of Food and Agriculture. doi:10.1002/jsfa.8818
166. Khodakovskaya, M., Dervishi, E., Mahmood, M., Xu, Y., Li, Z., Watanabe, F., & Biris, A. S. (2009, 09). Carbon Nanotubes Can Penetrate Plant Seed Coat and Dramatically Affect Seed Germination and Plant Growth. ACS Nano,3(10), 3221-3227. doi:10.1021/nn900887m
167. Khodakovskaya, M. V., Kole, C., & Kumar, D. S. (2016). Plant nanotechnology: Principles and practices. S. l.: Springer.
168. Kole, C., Kole, P., Randunu, K. M., Choudhary, P., Podila, R., Ke, P. C., . . . Marcus, R. K. (2013, 04). Nanobiotechnology can boost crop production and quality: First evidence from increased plant biomass, fruit yield, and phytomedicine content in bitter melon (Momordica charantia). BMC Biotechnology,13(1). doi:10.1186/1472-6750-13-37
169. Le, V. T., Bach, L. G., Pham, T. T., Le, N. T., Ngoc, U. T., Tran, D. N., & Nguyen, D. H. (2019, 03). Synthesis and antifungal activity of chitosan-silver nanocomposite synergize fungicide against Phytophthora capsici. Journal of Macromolecular Science, Part A,56(6), 522-528. doi:10.1080/10601325.2019.1586439
170. Lyons, K. (2010, January 04). Nanotechnology: Transforming Food and the Environment. Retrieved September 16, 2020, from https://foodfirst.org/publication/nanotechnology-transforming-food-and-the-environment/
171. Mercurio, M., Sarkar, B., & Langella, A. (2019). Modified clay and zeolite nanocomposite materials: Environmental and pharmaceutical applications. Amsterdam: Elsevier.
172. Nuruzzaman, M., Rahman, M. M., Liu, Y., & Naidu, R. (2016, 02). Nanoencapsulation, Nano-guard for Pesticides: A New Window for

Safe Application. Journal of Agricultural and Food Chemistry,64(7), 1447-1483. doi:10.1021/acs.jafc.5b05214

173. Onaga, G., & Wydra, K. (2016, 07). Advances in Plant Tolerance to Biotic Stresses. Plant Genomics. doi:10.5772/64351

174. Panpatte, D. G., Jhala, Y. K., Shelat, H. N., & Vyas, R. V. (2016). Nanoparticles: The Next Generation Technology for Sustainable Agriculture. Microbial Inoculants in Sustainable Agricultural Productivity, 289-300. doi:10.1007/978-81-322-2644-4_18

175. Shojaei, T. R., Salleh, M. A., Tabatabaei, M., Mobli, H., Aghbashlo, M., Rashid, S. A., & Tan, T. (2019). Applications of Nanotechnology and Carbon Nanoparticles in Agriculture. Synthesis, Technology, and Applications of Carbon Nanomaterials, 247-277. doi:10.1016/b978-0-12-815757-2.00011-5

176. Tripathi, D. K., Singh, S., Singh, V. P., Prasad, S. M., Dubey, N. K., & Chauhan, D. K. (2020, 09). Corrigendum to "Silicon nanoparticles more effectively alleviated UV-B stress than silicon in wheat (Triticum aestivum) seedlings" [Plant Physiol. Biochem. 110 (January 2017) 70-81]. Plant Physiology and Biochemistry,154, 796-797. doi:10.1016/j.plaphy.2020.05.027

177. Wang, S., Wang, F., & Gao, S. (2014, 09). Foliar application with nano-silicon alleviates Cd toxicity in rice seedlings. Environmental Science and Pollution Research,22(4), 2837-2845. doi:10.1007/s11356-014-3525-0

178. Afsharinejad, A., Davy, A., Jennings, B., & Brennan, C. (2016, 02). Performance Analysis of Plant Monitoring Nanosensor Networks at THz Frequencies. IEEE Internet of Things Journal,3(1), 59-69. doi:10.1109/jiot.2015.2463685

179. Chen, Y. W., Lee, H. V., Juan, J. C., & Phang, S. (2016, 10). Production of new cellulose nanomaterial from red algae marine biomass Gelidium elegans. Carbohydrate Polymers,151, 1210-1219. doi:10.1016/j.carbpol.2016.06.083

180. Gogos, A., Knauer, K., & Bucheli, T. D. (2012, 09). Nanomaterials in Plant Protection and Fertilization: Current State, Foreseen Applications, and Research Priorities. Journal of Agricultural and Food Chemistry,60(39), 9781-9792. doi:10.1021/jf302154y

181. Jampílek, J., & Kráľová, K. (2015, 09). Application Of Nanotechnology In Agriculture And Food Industry, Its Prospects And Risks. Ecological

Chemistry and Engineering S,22(3), 321-361. doi:10.1515/feces-2015-0018

182. Joga, M. R., Zotti, M. J., Smagghe, G., & Christiaens, O. (2016, 11). RNAi Efficiency, Systemic Properties, and Novel Delivery Methods for Pest Insect Control: What We Know So Far. Frontiers in Physiology,7. doi:10.3389/fphys.2016.00553
183. Khan, M. R., & Rizvi, T. F. (2014, 03). Nanotechnology: Scope and Application in Plant Disease Management. Plant Pathology Journal,13(3), 214-231. doi:10.3923/ppj.2014.214.231
184. Kitching, M., Ramani, M., & Marsili, E. (2014, 08). Fungal biosynthesis of gold nanoparticles: Mechanism and scale-up. Microbial Biotechnology,8(6), 904-917. doi:10.1111/1751-7915.12151
185. Kou, T., Yu, W., Lam, S. K., Chen, D., Hou, Y., & Li, Z. (2017, 12). Differential root responses in two cultivars of winter wheat (Triticum aestivum L.) to elevated ozone concentration under fully open-air field conditions. Journal of Agronomy and Crop Science,204(3), 325-332. doi:10.1111/jac.12257
186. Lee, W., Kang, S. H., Kim, J., Kolekar, G. B., Sung, Y., & Han, S. (2009, 07). TiO2 nanotubes with a ZnO thin energy barrier for improved current efficiency of CdSe quantum-dot-sensitized solar cells. Nanotechnology,20(33), 335706. doi:10.1088/0957-4484/20/33/335706
187. Lodewijks, K., Miljkovic, V., Massiot, I., Mekonnen, A., Verre, R., Olsson, E., & Dmitriev, A. (2016, 06). Multiscale conformal pattern transfer. Scientific Reports,6(1). doi:10.1038/srep28490
188. Lv, M., Liu, Y., Geng, J., Kou, X., Xin, Z., & Yang, D. (2018, 05). Engineering nanomaterials-based biosensors for food safety detection. Biosensors and Bioelectronics,106, 122-128. doi:10.1016/j.bios.2018.01.049
189. Mout, R., Ray, M., Tonga, G. Y., Lee, Y., Tay, T., Sasaki, K., & Rotello, V. M. (2017, 01). Direct Cytosolic Delivery of CRISPR/Cas9-Ribonucleoprotein for Efficient Gene Editing. ACS Nano,11(3), 2452-2458. doi:10.1021/acsnano.6b07600
190. Pradhan, D., Sindhwani, S., & Leung, K. T. (2010, 07). Parametric Study on Dimensional Control of ZnO Nanowalls and Nanowires by Electrochemical Deposition. Nanoscale Research Letters,5(11), 1727-1736. doi:10.1007/s11671-010-9702-2

191. Ran, Y., Liang, Z., & Gao, C. (2017, 05). Current and future editing reagent delivery systems for plant genome editing. Science China Life Sciences,60(5), 490-505. doi:10.1007/s11427-017-9022-1

192. Singh, D. P., Singh, H. B., & Prabha, R. (2016). Microbial inoculants in sustainable agricultural productivity. New Delhi: Springer.

193. Tiwari, J. N., Tiwari, R. N., & Kim, K. S. (2012, 05). Zero-dimensional, one-dimensional, two-dimensional, and three-dimensional nanostructured materials for advanced electrochemical energy devices. Progress in Materials Science,57(4), 724-803. doi:10.1016/j.pmatsci.2011.08.003

194. Torney, F., Trewyn, B. G., Lin, V. S., & Wang, K. (2007, 04). Mesoporous silica nanoparticles deliver DNA and chemicals into plants. Nature Nanotechnology,2(5), 295-300. doi:10.1038/nnano.2007.108

195. Vermeulen, S., Aggarwal, P., Ainslie, A., Angelone, C., Campbell, B., Challinor, A., . . . Wollenberg, E. (2012, 01). Options for support to agriculture and food security under climate change. Environmental Science & Policy,15(1), 136-144. doi:10.1016/j.envsci.2011.09.003

196. Vijayakumar, P. S., Abhilash, O. U., Khan, B. M., & Prasad, B. L. (2010, 06). Nanogold-Loaded Sharp-Edged Carbon Bullets as Plant-Gene Carriers. Advanced Functional Materials,20(15), 2416-2423. doi:10.1002/adfm.200901883

197. Yang, H., Xu, M., Koide, R. T., Liu, Q., Dai, Y., Liu, L., & Bian, X. (2015, 04). Effects of ditch-buried straw return on water percolation, nitrogen leaching, and crop yields in a rice-wheat rotation system. Journal of the Science of Food and Agriculture,96(4), 1141-1149. doi:10.1002/jsfa.7196

198. (n.d.). Retrieved September 16, 2020, from https://www.ars.usda.gov/research/publications/publication/?seqNo115=325896

199. Abbar, S., Schilling, M. W., Whitworth, R. J., & Phillips, T. W. (2016, 04). Efficacy of selected pesticides against Tyrophagus putrescentiae (Schrank): Influences of applied concentration, application substrate, and residual activity over time. Journal of Pest Science,90(1), 379-387. doi:10.1007/s10340-016-0766-3

200. Alarcón-Moyano, J., & Matiacevich, S. (2019, 01). Active emulsions based on alginate and lemongrass/citral essential oils: Effect of encapsulating agents on physical and antimicrobial properties. International Journal of Food Properties,22(1), 1952-1965. doi:10.1080/10942912.2019.1698605

201. Anjali, C., Sharma, Y., Mukherjee, A., & Chandrasekaran, N. (2011, 07). Neem oil (Azadirachta Indica) nanoemulsion-a potent larvicidal agent against Culex quinquefasciatus. Pest Management Science,68(2), 158-163. doi:10.1002/ps.2233
202. Anton, N., & Vandamme, T. F. (2010, 11). Nano-emulsions and Micro-emulsions: Clarifications of the Critical Differences. Pharmaceutical Research,28(5), 978-985. doi:10.1007/s11095-010-0309-1
203. Aswathanarayan, J. B., & Vittal, R. R. (2019, 11). Nanoemulsions and Their Potential Applications in Food Industry. Frontiers in Sustainable Food Systems,3. doi:10.3389/fsufs.2019.00095
204. Bera, A., & Mandal, A. (2014, 09). Microemulsions: A novel approach to enhanced oil recovery: A review. Journal of Petroleum Exploration and Production Technology,5(3), 255-268. doi:10.1007/s13202-014-0139-5
205. Chang, Y., & Mcclements, D. J. (2014, 03). Optimization of Orange Oil Nanoemulsion Formation by Isothermal Low-Energy Methods: Influence of the Oil Phase, Surfactant, and Temperature. Journal of Agricultural and Food Chemistry,62(10), 2306-2312. doi:10.1021/jf500160y
206. Donsì, F., Annunziata, M., Vincensi, M., & Ferrari, G. (2012, 06). Design of nanoemulsion-based delivery systems of natural antimicrobials: Effect of the emulsifier. Journal of Biotechnology,159(4), 342-350. doi:10.1016/j.jbiotec.2011.07.001
207. Gaysinsky, S., Taylor, T. M., Davidson, P. M., Bruce, B. D., & Weiss, J. (2007, 11). Antimicrobial Efficacy of Eugenol Microemulsions in Milk against Listeria monocytogenes and Escherichia coli O157:H7. Journal of Food Protection,70(11), 2631-2637. doi:10.4315/0362-028x-70.11.2631
208. Hoar, T. P., & Schulman, J. H. (1943, 07). Transparent Water-in-Oil Dispersions: The Oleopathic Hydro-Micelle. Nature,152(3847), 102-103. doi:10.1038/152102a0
209. Isman, M. B. (2006, 01). Botanical Insecticides, Deterrents, And Repellents In Modern Agriculture And An Increasingly Regulated World. Annual Review of Entomology,51(1), 45-66. doi:10.1146/annual.into.51.110104.151146
210. Lam, A. C., & Schechter, R. S. (1987, 11). The theory of diffusion in microemulsion. Journal of Colloid and Interface Science,120(1), 56-63. doi:10.1016/0021-9797(87)90322-5

211. Lawrence, M. J. (1994, 09). Surfactant systems: Microemulsions and vesicles as vehicles for drug delivery. European Journal of Drug Metabolism and Pharmacokinetics,19(3), 257-269. doi:10.1007/bf03188929

212. Natural Remedies in the Fight Against Parasites. (2017, 07). doi:10.5772/63275

213. Papanikolaou, N. E., Kalaitzaki, A., Karamaouna, F., Michaelakis, A., Papadimitriou, V., Dourtoglou, V., & Papachristos, D. P. (2017, 02). Nano-formulation enhances the insecticidal activity of natural pyrethrins against Aphis gossypii (Hemiptera: Aphididae) and retains their harmless effect on non-target predators. Environmental Science and Pollution Research,25(11), 10243-10249. doi:10.1007/s11356-017-8596-2

214. Pavoni, L., Benelli, G., Maggi, F., & Bonacucina, G. (2019). Green nanoemulsion interventions for biopesticide formulations. Nano-Biopesticides Today and Future Perspectives, 133-160. doi:10.1016/b978-0-12-815829-6.00005-x

215. Perlatti, B., Bergo, P. L., Maria Fatima Das Gracas Fernandes Da Silva, Batista, J., & Rossi, M. (2013, 01). Polymeric Nanoparticle-Based Insecticides: A Controlled Release Purpose for Agrochemicals. Insecticides - Development of Safer and More Effective Technologies. doi:10.5772/53355

216. Sasson, Y., Levy-Ruso, G., Toledano, O., & Ishaaya, I. (2007). Nanosuspensions: Emerging Novel Agrochemical Formulations. Insecticides Design Using Advanced Technologies, 1-39. doi:10.1007/978-3-540-46907-0_1

217. Schulman, J. H., Stoeckenius, W., & Prince, L. M. (1959, 10). Mechanism of Formation and Structure of Micro Emulsions by Electron Microscopy. The Journal of Physical Chemistry,63(10), 1677-1680. doi:10.1021/j150580a027

218. Shah, D. O. (1998). Micelles, microemulsions, and monolayers: Science and technology. New York: Dekker.

219. Tadros, T., Izquierdo, P., Esquena, J., & Solans, C. (2004, 05). Formation and stability of nano-emulsions. Advances in Colloid and Interface Science,108-109, 303-318. doi:10.1016/j.cis.2003.10.023

220. Talegaonkar, S., Azeem, A., Ahmad, F., Khar, R., Pathan, S., & Khan, Z. (2008, 11). Microemulsions: A Novel Approach to Enhanced Drug

Delivery. Recent Patents on Drug Delivery & Formulation,2(3), 238-257. doi:10.2174/187221108786241679

221. Terjung, N., Löffler, M., Gibis, M., Hinrichs, J., & Weiss, J. (2012). Influence of droplet size on the efficacy of oil-in-water emulsions loaded with phenolic antimicrobials. Food Funct.,3(3), 290-301. doi:10.1039/c2fo10198j
222. Venhuis, S. H., & Mehrvar, M. (2004). Health effects, environmental impacts, and photochemical degradation of selected surfactants in water. International Journal of Photoenergy,6(3), 115-125. doi:10.1155/s1110662x04000157
223. Wilhelm, K., Cua, A. B., Wolff, H. H., & Maibach, H. I. (1993, 09). Surfactant-Induced Stratum Corneum Hydration InVivo: Prediction of the Irritation Potential of Anionic Surfactants. Journal of Investigative Dermatology,101(3), 310-315. doi:10.1111/1523-1747.ep12365467
224. Campbell, B. M., Thornton, P., Zougmoré, R., Asten, P. V., & Lipper, L. (2014, 10). Sustainable intensification: What is its role in climate-smart agriculture? Current Opinion in Environmental Sustainability,8, 39-43. doi:10.1016/j.cosust.2014.07.002
225. Chahine, N. O., Collette, N. M., Thomas, C. B., Genetos, D. C., & Loots, G. G. (2014, 09). Nanocomposite Scaffold for Chondrocyte Growth and Cartilage Tissue Engineering: Effects of Carbon Nanotube Surface Functionalization. Tissue Engineering Part A,20(17-18), 2305-2315. doi:10.1089/ten.tea.2013.0328
226. Kah, M., & Hofmann, T. (2014, 02). Nanopesticide research: Current trends and future priorities. Environment International,63, 224-235. doi:10.1016/j.envint.2013.11.015
227. Nath, R., García, R. E., Blendell, J. E., & Huey, B. D. (2007, 01). The influence of grain boundaries and texture on ferroelectric domain hysteresis. Jom,59(1), 17-21. doi:10.1007/s11837-007-0004-9
228. Oliveira, J. L., Campos, E. V., Bakshi, M., Abhilash, P., & Fraceto, L. F. (2014, 12). Application of nanotechnology for the encapsulation of botanical insecticides for sustainable agriculture: Prospects and promises. Biotechnology Advances,32(8), 1550-1561. doi:10.1016/j.biotechadv.2014.10.010
229. Prasad, R., Bhattacharyya, A., & Nguyen, Q. D. (2017, 06). Nanotechnology in Sustainable Agriculture: Recent Developments, Challenges, and Perspectives. Frontiers in Microbiology,8. doi:10.3389/fmicb.2017.01014

230. Rajendran, S. P., & Sengodan, K. (2017). Synthesis and Characterization of Zinc Oxide and Iron Oxide Nanoparticles Using Sesbania grandiflora Leaf Extract as Reducing Agent. Journal of Nanoscience,2017, 1-7. doi:10.1155/2017/8348507

231. Sekhon, B. (2014, 05). Nanotechnology in agri-food production: An overview. Nanotechnology, Science and Applications, 31. doi:10.2147/NSA.s39406

232. Sertova, N. (2015). Application of nanotechnology in the detection of mycotoxins and the agricultural sector. Journal of Central European Agriculture,16(2), 117-130. doi:10.5513/jcea01/16.2.1597

233. Sun, C., Shu, K., Wang, W., Ye, Z., Liu, T., Gao, Y., . . . Yin, Y. (2014, 03). Encapsulation and controlled release of hydrophilic pesticide in shell cross-linked nanocapsules containing aqueous core. International Journal of Pharmaceutics,463(1), 108-114. doi:10.1016/j.ijpharm.2013.12.050

234. Vidotti, M., Carvalhal, R. F., Mendes, R. K., Ferreira, D. C., & Kubota, L. T. (2011, 01). Biosensors based on gold nanostructures. Journal of the Brazilian Chemical Society,22(1), 3-20. doi:10.1590/s0103-50532011000100002

235. Yunlong, C., & Smit, B. (1994, 07). Sustainability in agriculture: A general review. Agriculture, Ecosystems & Environment,49(3), 299-307. doi:10.1016/0167-8809(94)90059-0

236. Yunlong, C., & Smit, B. (1994, 07). Sustainability in agriculture: A general review. Agriculture, Ecosystems & Environment,49(3), 299-307. doi:10.1016/0167-8809(94)90059-0

237. Jefferson, R. A. (1993, 12). Beyond Model Systems. New Strategies, Methods, and Mechanisms for Agricultural Research. Annals of the New York Academy of Sciences,700(1 Biotechnology), 53-73. doi:10.1111/j.1749-6632.1993.tb26306.x

238. Krizek, B. A., Prost, V., Joshi, R. M., Stoming, T., & Glenn, T. C. (2003, 01). Developing transgenic Arabidopsis plants to be metal-specific bioindicators. Environmental Toxicology and Chemistry,22(1), 175-181. doi:10.1002/etc.5620220123

239. Mehrotra, R., Gupta, G., Sethi, R., Bhalothia, P., Kumar, N., & Mehrotra, S. (2011, 02). Designer promoter: An artwork of cis engineering. Plant Molecular Biology,75(6), 527-536. doi:10.1007/s11103-011-9755-3

240. Tsien, R. Y. (1998, 06). The Green Fluorescent Protein. Annual Review of Biochemistry,67(1), 509-544. doi:10.1146/annual.biochem.67.1.509

INDEX

A

Abiotic interactions 55
Abiotic stresses 14, 15
Abnormal brain development 82
Abundance of the root hairs 81
Accurate method 126
Acidification 105, 106
Adsorption reactions 34
Aeration process 31
Ag nanoparticles 159
Agricultural crops 72
Agricultural ecosystems and natural systems 33
Agricultural growth 73
Agricultural nanotechnology 159, 176
Agricultural practices 73, 150, 178, 185
Agricultural production systems 98
Agricultural soils 4
Agricultural sustainability 139
Agroecology studies 55
Agroecosystem 158
Agroecosystems 11
Agronomic efficiency 32, 33
Alkaline soils 140, 143, 145, 146, 147, 149
Alternative energy 72
Ammonia fixation 32
Ammonification 31
Anion exchange 107
Anthropogenic intervention 56
Arabidopsis plant 28
Arabidopsis plants 76
Arabidopsis species 99
Ascorbic acid oxidase 43
Assimilation 13
Atmospheric nitrogen fixation 4
Atmospheric pressure chemical ionization or atmospheric pressure photon ionization (APCI/APPI) 123
Auxin concentration 101
Azospirillum species 74, 77, 81, 93

B

Bacterial siderophores 107

Bacterial species 74, 75, 76, 77, 80, 81, 82, 84, 90, 92, 93
Bicarbonate concentration 141, 147
Bifunctional enzyme 119
Biofertilizers 73
Biomass production 98
Breeding mechanism 9

C

Calcareous soils 144, 145, 147, 148, 149
Calcium deficiency 37
Calvin cycle 120
Capillary electrophoresis 123, 126, 132
Cell division 80, 95
Chemical fertilizers 31, 35, 39, 44, 49, 72, 73, 74, 76, 82
Chloroplastic capacity 109
Climate change 8
Climatic conditions 6
Crop classification process 7
Crop genotypes 107
Crop infestation 43
Crop nutrient efficiency 32
Crop plants 6, 7
Crop production 56, 98
Crop productivity 4, 11
Crop recycling 42
Crop rotation 83
Crop rotation process 140, 143
Crop species 4, 6
Crop yields 4, 6, 8, 9, 11, 12, 14, 21, 22
Cyanobacteria 84
Cyclotron frequency 124
Cytochrome oxidase 43
Cytokinin activity 80

D

Deep-rooted plants 104
Denitrification process 32, 106
Detoxification 13
Diffusion coefficient 35
Dinitrogenase component 78
Dinitrogenase reductase 78
Direct developmental process 55
Divalent cation 36
Dolomite 142
Dynamic system 31

E

Ecological farming 154
Ecotoxicity 160
Elementary flux models 119
Enzyme activation process 38
Essential nutrients 5, 10, 13
Ethylene production 67
Eutrophication 6
Exhibit symbioses 101

F

Fusarium oxysporium 74

G

Gas chromatography 123, 125
Genetic engineering 109, 113, 114
Genetic variation 102
Genetic variation complexity 59
Genome-wide techniques 61
Genotypes 52, 53, 58, 63
Genotypic performance 12
Global warming 109
Glycine molecules 120
Glycolate molecules 120, 122
Grain production 4
Gravitropic sensitivity 101

H

Healing process 36
High molecular weight 103
High nutrient concentration 101
Homogenous magnetic field 124
Hydrogenase enzyme 78
Hydrolysis reactions 78

I

Immunity disorders 82
Inorganic nutrients 103
Integrated nutrient management 140, 141, 142, 144, 153, 154
Intensive cultivation 138
Intermediate metabolites 26
Iron deficiency 105
Irregular drainage 28

L

Laser ablation electrospray ionization technique 131
Leaching process 37
Legume-grass mixture 13
Leguminosae 7

M

Macronutrients 4, 6, 10, 13
Macrophomina phaseolina 73, 74
Mass spectroscopy 123, 124
Matrix-assisted laser desorption/ionization 123
Meristematic tissues 34
Mesophyll cells 109
Metabolic absorption 4
Metabolic enzymes 8
Metabolic flux 119, 130, 133
Metabolism 26, 27, 28, 29, 42, 44, 45, 47, 49
Metabolomic repository 131
Methylotrophic bacteria 80
Microbial inoculants 73, 75
Microbial solubilization 108
Micronutrient absorption 13
Micronutrient deficiency 4
Micronutrients 4, 6, 10, 14
Micropropagation 26
Mineral fertilizers 52
Mineralization 105, 106, 108
Mineralization process 56
Mineral nutrients 98, 108
Mineral nutrition 4
Mycorrhizal fungi 107, 108

N

Nano fertilizers 158, 160, 161, 162, 163, 169, 173, 174, 175, 176, 183
Nanonutrients 158
Nanopesticides 158
Nanotechnology 154
Near Isogenic lines (NILs) 59
Nitrification process 31, 47
Nitrogenase enzyme 78
Nitrogen fixation 27, 31, 36, 43, 45, 75, 78, 91, 92, 106
Nutrient adaptability 13
Nutrient capacity 56
Nutrient concentration 55
Nutrient concentrations 9, 10, 16
Nutrient deficiency 33
Nutrient management 72, 74
Nutrient mobilization 103
Nutrient-rich environments 56
Nutrient toxicity 11, 13
Nutrient uptake mechanisms 7, 9, 13
Nutrient utilization efficiency 9
Nutritional deficiencies 6

Nutritional quality 30

O

Organic matter 31, 38, 39, 40, 41, 42, 43, 44, 48, 49
Organic products 98
Osmotic stress 80, 93

P

Pale green foliage 141, 142, 148
Pericycle cells 101
Phenolic compounds 107
Phenotypic variation 57, 58
Phosphate solubilizing microbial cultures 141
Phosphobacterium 74
Phosphorous deficiency 33
Photorespiration 119, 120, 121
Photorespiration process 43
Photosynthesis process 29, 32, 109, 121
Photosynthetic efficiency 14
Physiological efficiency 32, 33
Physiological processes 73, 80, 94
Phytin component 34
Phytonanotechnology 158
Phytopathogenic microbes 75
Plant breeding techniques 8, 9
Plant cells 5
Plant-derived chelators 107
Plant genotype 103
Plant growth 99
Plant growth-promoting bacteria 75
Plant growth-promoting rhizobacteria 73
Plant growth regulators 28
Plant hormones 68
Plant metabolic activities 54
Plant metabolic networks 119
Plant nutrition 72
Plant systems 55, 56, 57, 64, 65, 67, 69
Plasma membrane 26, 28
Polyacrylic acid 160
Productive carboxylation process 119
Pyruvate molecules 121, 122
Pythium debaryanum 73

Q

Quadrupole mass analyzer 124
Quantitative trait loci (QTLs) 58

R

Radiation interception 108, 109
Radioactive nuclides 27
Rapid nutrient absorption 13
Rapid oxidation 105
Recombinant inbred lines 59
Redox potential 105
Respiration process 105
Retention capacity 36
Rhizobium leguminosarum 78
Rhizobium population 78
Rhizodeposition 101
Rhizosphere microbes 107
Rhizosphere region 101, 107
Rhizosphere zone 103, 104, 105, 106
Ribulose Bisphosphate carboxylase oxygenase 119
Root architecture 100, 102
Root geometry 102
Root growth 101, 103, 108, 113
Root penetration problems 139
Root system 99, 100, 102, 106
Root trait 98

S

Saline soils 144, 147
Senescence mechanism 109, 110
Senescence process 109, 110, 111
Signal transduction pathway 101
Sodic soils 141, 144, 147
Soil biology 6
Soil erosion 6, 11, 22
Soil fertility 138, 139, 149, 154
Soil moisture 52
Soil organic matter 52
Soil water composition 36
Solubilization process 75, 85
Sorghum crop 149
Sorghum plants 146
Sorghum species 146
Stomatal absorption 57
Sugarcane yield 9
Systematic improvement 154

T

Thermal ionization 123
Tissue culture 26
Tomato crops 160
Transamination process 120
Transformation process 31
Translocation 13
Transporter molecules 119
Transporter proteins 26, 28, 29

W

Waterlogged conditions 140, 146
Water molecules 104

X

Xanthomonas campestris 74

Y

Yellowing process 109

Z

Zinc deficiency 82, 83
Zinc resistance mechanisms 84

Printed in the United States
by Baker & Taylor Publisher Services